The Necessities Underlying Reality

Also available from Bloomsbury:

Knowledge and the Philosophy of Number by Keith Hossack
Fragmenting Reality by Samuele Iaquinto and Giuliano Torrengo
Aristotle's Syllogism and the Creation of Modern Logic by Lukas M. Verburgt

The Necessities Underlying Reality

Connecting Philosophy of Mathematics, Ethics and Probability

James Franklin

Edited by
Jeremiah Joven Joaquin & James Franklin

BLOOMSBURY ACADEMIC
LONDON • NEW YORK • OXFORD • NEW DELHI • SYDNEY

BLOOMSBURY ACADEMIC
Bloomsbury Publishing Plc, 50 Bedford Square, London, WC1B 3DP, UK
Bloomsbury Publishing Inc, 1385 Broadway, New York, NY 10018, USA
Bloomsbury Publishing Ireland, 29 Earlsfort Terrace, Dublin 2, D02 AY28, Ireland

BLOOMSBURY, BLOOMSBURY ACADEMIC and the Diana logo are trademarks of Bloomsbury Publishing Plc

First published in Great Britain 2026

A catalogue record for this book is available from the British Library.

A catalog record for this book is available from the Library of Congress.

ISBN: HB: 978-1-3504-6707-1
ePDF: 978-1-3504-6708-8
eBook: 978-1-3504-6709-5

Series: Expanding Philosophy of Religion

Typeset by RefineCatch Limited, Bungay, Suffolk

For product safety related questions contact productsafety@bloomsbury.com.

Contents

Figures

Acknowledgements

Chapter 5 contains reworked material from *What Science Knows: And How It Knows It* (Encounter Books, 2009) and Chapter 10 contains reworked material from *The Worth of Persons: The Foundation of Ethics* (Encounter Books, 2022). Permission of Encounter Books is acknowledged.

Springer Nature's permission is acknowledged for the use of reworked material from: 'Mathematics as a Science of Non-Abstract Reality: Aristotelian Realist Philosophies of Mathematics', *Foundations of Science* 27 (2022), 327–44, in Chapter 3; 'Global and Local', *Mathematical Intelligencer* 36, no. 4 (Dec 2014): 4–9, in Chapter 7; and '"Let No-One Ignorant of Geometry . . .": Mathematical Parallels for Understanding the Objectivity of Ethics', *The Journal of Value Inquiry* 57 (2023): 365–84, in Chapter 9.

Reworked material is used under Open Access licences from the following: 'Discrete and Continuous: A Fundamental Dichotomy in Mathematics', *Journal of Humanistic Mathematics* 7 (2017): 355–78, in Chapter 7; 'The Global-Local Distinction Vindicates Leibniz's Theodicy', *Theology and Science* 20(4) (2022), in Chapter 8.

A Celebratory Prologue

Jeremiah Joven Joaquin

This book celebrates the life and work of one of my dear philosophical mentors, James Franklin. Jim had a stellar professional career at the University of New South Wales, his academic home since 1981. When he retired in 2019, I proposed publishing a volume that encapsulated his views on the many topics he worked on throughout his academic tenure. This book is such a volume.

Jim was a professor at UNSW's School of Mathematics and Statistics. One may think that he is purely a mathematician. However, as a 2019 citation of the Royal Society of New South Wales evinces, he is not only a mathematician but also a "leading expert in philosophy, ethics, and the history of ideas."[1] Indeed, his body of work covers a broad range of topics, from philosophical issues about the nature of science and mathematics to more practical questions about the ethics of drug testing welfare recipients and the morality of immigration. His *What Science Knows: And How It Knows It* (2009) and *An Aristotelian Realist Philosophy of Mathematics* (2014) are instances of the former sort, while his *Catholic Values and Australian Realities* (Connor Court Publishing, 2006) is of the latter kind. His research in the history of ideas is evident in works such as *The Science of Conjecture: Evidence and Probability before Pascal* (2001; revised edition, 2015) and *Arthur Calwell: Transforming Australia through Immigration* (2023).[2]

It was a chance reading of one of his history of ideas books that led to my friendship with Jim. I was fresh out of a PhD fellowship at the Australian National University when I got hold of his *Corrupting the Youth: A History of Philosophy in Australia* (2003) in late 2010. It was lent to me by Brenton Welford, a good friend from Monash University. *Corrupting the Youth* presents a witty and judicious but highly opinionated, Sydney-centric, and polemical history of philosophy Down Under. While it had its share of critics,[3] it was well-received. As David Armstrong – one of the more prominent Australian philosophers – attests, it is "an important contribution to the cultural history of our country."

So, what is distinctive about Australian philosophy? According to Jim's *Corrupting the Youth*, Australian philosophy is a realist philosophy – realist 'in the sense of being concerned with the ways of working of real things in the world rather than having our ideas as the central focus of philosophy' (p. 3).

Most Australian philosophers are indeed realists about many things. Some are realists about scientific entities; others are realists about minds. Others (mostly Catholic Australian philosophers like Jim) are realists about ethical truths, while others are realists about universals and laws of nature.

That Jim has this view about the stamp of Australian philosophy is unsurprising given that he was influenced by Armstrong and David Stove, two of the most vocal proponents of Australian realism, and his teachers at the University of Sydney in the early 1970s. As many professional philosophers would know, Armstrong, along with J. J. C. Smart and U. T. Place, are best known for their materialist theory of mind,[4] while Stove is best known for his polemics against the irrationalists (Karl Popper, Paul Feyerabend, Thomas Kuhn, and Imre Lakatos).[5] Incidentally, Jim serves as Stove's literary executor.[6]

Jim and I officially met in May 2013. I invited him to give the keynote address at a conference I co-organized in Boracay (a glorious island south of Manila, Philippines). In the blazing summer heat, he delivered his views about the nature and importance of a realist philosophy. Jim and I must have hit it off during his visit as we have regularly corresponded and have had further collaborations thereafter.

In 2014, we planned to co-teach a philosophy of mathematics course at De La Salle University, my home university. This course was based on his *An Aristotelian Realist Philosophy of Mathematics*. In 2014 and 2015, he invited me to be a visiting researcher at UNSW to work on philosophical issues in logic and probability. These visits translated into a probability conference we co-organized at the Australian National University in 2015 and a co-authored paper, 'A Causal-Mentalist View of Propositions', published in *Organon F: International Journal of Analytic Philosophy* in 2022. In 2020, at the height of the Covid-19 lockdowns, we had a fruitful discussion on Armstrong, truthmaking, and Australian philosophy on my YouTube channel, *Philosophy and What Matters*.[7]

This book is another product of my decade-long collaboration and friendship with Jim. In its pages, a reader will find Jim's philosophical reflections about the necessities found in mathematics, science, ethics and the history of ideas. These necessities are not just features of our language or conceptual scheme. They are rather real, mind- and language-independent features of the world. In a more metaphorical parlance, these absolute necessities carve reality at its natural joints.

In this sense, Jim's view that absolute necessities are real features of the world is Aristotelian, not Platonist or nominalist. Since such necessities are found in the world, they are not found in some other-worldly realm nor in the structure of human language or thought. On this view, mathematical and ethical necessities exist in the physical world we live in.

One may wonder how Jim arrived at this conclusion. How are absolute necessities real features of the world? Moreover, what do the necessities found in disparate areas like the abstract area of mathematics and the practical realm of ethics have in common? For instance, what does the mathematical truth that multiplication is commutative have to do with the ethical truth, if it is even a truth, that drug testing welfare recipients is morally wrong?

As the subsequent chapters will show, Jim employs one basic line of argument for his Aristotelian view about the absolute necessities found in mathematics, science, ethics, and the history of ideas, viz., a kind of inference to the best explanation (though he adds that the simpler necessities are directly accessible and do not need any inference to reach them). Compared to other views, viz., Platonist and nominalist, Aristotelian realism provides the best and most coherent account of the objectivity and the 'contentfulness' of such necessities. This means that mathematical and ethical necessities, if they are truly absolute necessities, share the same objective, mind- and language-independent nature. They only differ in their contents: mathematical necessities are about mathematical facts found in the world; ethical necessities are about ethical facts grounded in our worth as human persons.

Plan of the Book

This book is divided into six main parts. Part I presents Jim's life and work as an engaged search for absolute necessities. His intellectual autobiography (Chapter 1) describes the very unusual path needed in the present day to reach a clear view of absolute necessities. His lifelong study of genuine necessities in mathematics was combined with an exposure to Aristotelian realist ideas in philosophy, which provided a way to interpret those necessities. Contrary to the sceptical and deconstructionist spirit of the age, such a training reconnects with older views, such as in Aristotle's *Posterior Analytics*, that the world is full of necessary connections.

Chapter 2 presents some basic examples of such necessities, starting with those as simple as 'Orange is between red and yellow'. It then shows how current philosophy, cut off as it is from both mathematics and the Aristotelian philosophical heritage, has failed to give an account of them.

The rest of the book gives that account, across three different areas – mathematics (including applied mathematics), logic (both probabilistic and deductive) and ethics. It concludes with an historical section, explaining how absolute necessities came to be understood over many centuries, but how our view of them was then obscured in modern times.

Part II deals with mathematics, the home ground of necessities. For twenty-three centuries Euclid's *Elements* and its successors have provided training in the proof of mathematical certainties from self-evident axioms. An Aristotelian realist philosophy of mathematics (Chapter 3) – in contrast to nominalist and Platonist theories – interprets that correctly, showing that mathematics is a true science of the world, dealing with its quantitative and structural properties (such as, respectively, ratio and symmetry). However, it is not an empirical science like biology, but it can establish necessary truths about those properties and relations (as in the examples at the beginning of Chapter 2).

In more complex cases, mathematics utilizes the method of deductive proof (Chapter 4); this allows the generation of an infinite number of certain truths about necessities from simple ones. In recent times, the method of proof has extended beyond mathematics to the 'formal sciences' such as computing and the interconnections of the laws of physics (though the individual laws of physics remain unprovable).

Part III deals with the absolute necessities of logic, both probabilistic and deductive. The relation of uncertain evidence to conclusion (in courts of law and experimental science, for example) is a purely logical one (Chapter 5). For example, one could not start experimental science without the principle that the predictions of a theory confirm it (make it objectively more credible), rather than telling against it; the principle itself is logical and not revisable. That principle is set in a theory of 'logical probability' or 'objective Bayesianism' which gives an overall account of the nature of the relation of evidence to conclusion.

Deductive logic (Chapter 6) is also a science of objective real-world necessities – or at least is founded on them – rather than being about tautologies, or language, or thought, or conventions. While there can be trivial logical necessities such as 'If *p* then *p*', more substantial ones express an inherent 'pre-logical' necessary structure of the world as consisting of entities having properties and relations, as described in Aristotle's *Categories*.

Part IV develops two very general themes of mathematics, discrete/continuous and local/global (Chapter 7) which are 'well-known' but rarely discussed and hard to explain without an Aristotelian framework. The counting numbers are discrete while quantities like length and time are continuous, and discrete mathematics (such as number theory and algebra) is very different from continuous (such as calculus); yet there are necessary connections between the two, like the incommensurability of the diagonal and side of a square.

Similar connections exist between what is possible locally (for example to walk over a few of the Königsberg bridges just once) and what is possible, or impossible, globally (to walk over all of them just once). Since such necessities are absolute, they are not subject to exception by God. That has significance in supporting Leibniz's solution to the problem of evil in the philosophy of religion (Chapter 8). His answer to the question 'Why does God permit terrible evil?' is that he cannot – because this is already the best of all possible worlds, in that trying to improve it locally (which we naively imagine would make it better overall) would make it worse globally.

Part V deals with the necessities of ethics. Ethics parallels mathematics (but not natural science), in being a realm in which necessities are accessible by pure thought (Chapter 9). The prominent role of equality in both is not an accident but reflects insights into, for example, the equal worth of persons.

Ethics in the sense of what to do depends on something deeper (Chapter 10), the properties of humans that make the death of a human a tragedy but the explosion of a lifeless galaxy merely a firework. The inherent worth of persons – what makes them matter and gives them rights and duties – is an absolutely necessary consequence of (supervenes on) their unique properties, such as rationality, consciousness and emotional structure.

Part VI deals with the history of ideas on a large scale, to show the long-term story of the discovery of absolute necessities (Chapter 11). The Scientific Revolution was largely created not by Baconian scientists building up theories from patient observations, but by would-be Euclids like Galileo, Pascal and Newton aiming to establish physical necessities by mathematical means – with partial, though not total, success. Some of the social sciences, such as economics, were also heavily based on purely conceptual discoveries, which are necessary to organise the data.

Unfortunately, since the Enlightenment, the more necessities that have been discovered (for example in the golden age of mathematics in the twentieth century), the less philosophy and the general culture have been interested in them (Chapter 12). Hume and his successors in Anglophone philosophy, and various proponents of 'hermeneutics of suspicion' such as Marx and Freud, have undermined our confidence in absolute certainties. It is time for civilization to recover its knowledge of absolute necessities.

Chapter Sources

Chapters 1, 2 and 12 are published here for the first time. The other chapters are based on Jim's previously published work, though substantially rewritten for this book. We thank their original publishers for permitting us to use them here.

Chapter 3: Aristotelian Realist Philosophy of Mathematics. *An Aristotelian Realist Philosophy of Mathematics*; 'Mathematics as a Science of Non-Abstract Reality: Aristotelian Realist Philosophies of Mathematics'; *Foundations of Science* 27 (2022): 327–44; and, 'The Mathematical World', *Aeon* 7 Apr 2014, https://aeon.co/essays/aristotle-was-right-about-mathematics-after-all.

Chapter 4: Proof in Mathematics and Beyond. James Franklin and Albert Daoud, *Proof in Mathematics: An Introduction*, (2010), ch 2; *An Aristotelian Realist Philosophy of Mathematics*, chs 9 and 12; 'The Formal Sciences Discover the Philosophers' Stone', *Studies in History and Philosophy of Science* 25 (1994): 513–33; 'Mathematics, Core of the Past and Hope of the Future', in *Reclaiming Education: Renewing Schools and Universities in Contemporary Western Society*, ed. C. Runcie and D. Brooks (2018), 149–62; and, 'Early Modern Mathematical Principles and Symmetry Arguments', in *The Idea of Principles in Early Modern Thought*, ed. P. Anstey (2017), ch. 1.

Chapter 5: Logical Probability and the Relation of Evidence to Hypothesis. *What Science Knows: And How It Knows It*, chs 1 and 10; 'Resurrecting Logical Probability', *Erkenntnis* 55 (2001): 277–305; 'The Objective Bayesian Conceptualisation of Proof and Reference Class Problems', *Sydney Law Review* 33 (2011): 545–61; 'Non-deductive Logic in Mathematics', *British Journal for the Philosophy of Science* 38 (1987): 1–18, revised in 'Non-deductive Logic in Mathematics: The Probability of Conjectures', in *The Argument of Mathematics*, ed. A. Aberdein and I. Dove (2013), 11–29; Keith Campbell, James Franklin and Douglas Ehring, 'Donald Cary Williams', *Stanford Encyclopedia of Philosophy*, 2013/2023, sections 6–7.

Chapter 6: The Grounding of Logic in Prelogical Necessities. Jeremiah Joven Joaquin and James Franklin, 'A Causal-Mentalist View of Propositions', *Organon F: International Journal of Analytic Philosophy*, 29 (2022): 47–77, https://doi.org/10.31577/orgf.2022.29103.

Chapter 7: Two Great Themes of Mathematics: Discrete/Continuous and Local/Global. 'Discrete and Continuous: A Fundamental Dichotomy in Mathematics', *Journal of Humanistic Mathematics* 7 (2017): 355–78; 'Global and Local', *Mathematical Intelligencer* 36, no. 4 (Dec 2014): 4–9; 'The Global-local Distinction Vindicates Leibniz's Theodicy', *Theology and Science* 20, no. 4 (2022).

Chapter 8: Leibniz's Solution to the Problem of Evil via Absolute Constraints on God. 'The Global-local Distinction Vindicates Leibniz's Theodicy', *Theology and Science* 20 (4) (2022); 'Antitheodicy and the Grading of Theodicies by Moral Offensiveness', *Sophia* 59 (2020): 563–76; 'Two Caricatures II: Leibniz's Best World', *International Journal for Philosophy of Religion* 52 (2002): 45–56.

Chapter 9: Ethics: The Parallel with Mathematics. '"Let No-one Ignorant of Geometry . . .": Mathematical Parallels for Understanding the Objectivity of Ethics', *Journal of Value Inquiry* 57 (2023): 365–84; 'On the Parallel between Mathematics and Morals', *Philosophy* 79 (2004): 97–119.

Chapter 10: Ethics: The Foundation in the Worth of Persons. *The Worth of Persons: The Foundation of Ethics* (2022).

Chapter 11: An Alternative, Aristotelian, History of Science. *The Science of Conjecture: Evidence and Probability Before Pascal* (Baltimore: Johns Hopkins University Press, 2001/2015); 'Prehistory of Probability', in *The Oxford Handbook of Probability and Philosophy,* ed. A. Hájek and C. Hitchcock (Oxford: Oxford University Press, 2016), 33–49; 'Late Scholastic Probable Arguments and Their Contrast with Rhetorical and Demonstrative Arguments', *Philosophical Inquiries* 10, no. 2 (2022); 'Science by Conceptual Analysis: The Genius of the Late Scholastics', *Studia Neoaristotelica* 9 (2012): 3–24; 'Early Modern Mathematical Principles and Symmetry Arguments', in *The Idea of Principles in Early Modern Thought*, ed. Peter Anstey (2017), ch. 1; 'Artifice and the Natural World: Mathematics, Logic, Technology', in *Cambridge History of Eighteenth Century Philosophy*, ed. Knud Haakonssen (2006), 817–53; 'The Formal Sciences Discover the Philosophers' Stone', *Studies in History and Philosophy of Science* 25 (1994), 513–33, updated as ch 6 of *An Aristotelian Realist Philosophy of Mathematics.*

Manila, Philippines
8 January 2025

Notes

1 See https://www.royalsoc.org.au/fellows/james-franklin-fellow-2019-sydney/

2 Jim's body of work can be accessed through his official website: https://web.maths.unsw.edu.au/~jim/. Some of their historical background are discussed in Chapter 1: An Intellectual Autobiography.

3 See, e.g., C. A. J. Coady's 2006 review at the *Notre Dame Philosophical Reviews*, https://ndpr.nd.edu/reviews/corrupting-the-youth-a-history-of-philosophy-in-australia/.

4 D. M. Armstrong, *A Materialist Theory of the Mind* (New York: Routledge), 1968.

5 D. C. Stove, *Popper and After: Four Modern Irrationalists* (New York: Pergamon Press), 1982; reprinted as *Anything Goes: Origins of the Cult of Scientific Irrationalism* (Sydney, Australia: Macleay Press), 1998.

6 Jim maintains David Stove's official website: https://web.maths.unsw.edu.au/~jim/davidstove.html.

7 Our discussion can be viewed at https://www.youtube.com/watch?v=QO_uvEUac_M.

Part One

Introducing Absolute Necessities

1

An intellectual autobiography

I arrived at the University of Sydney in February 1971, aged 17, equipped with certain advantages, interests and prejudices, but without a coherent plan. Having been born in the 1950s white, male, intelligent, from a stable family, and (apparently at least) healthy, I had made an early start on winning first prize in the lottery of life. My school education in various Catholic schools was mostly of high quality. In the last two years of school, I was fortunate to win a scholarship to St Joseph's College, Hunters Hill, where I enjoyed excellent teaching especially in mathematics (from Br Kevin Friel) and Latin (Br Michael Naughtin). For later developments, two elements of what was taught were especially significant. One was mathematical reasoning, including an adequate introduction to proof in geometry in the style of Euclid. The other was the 'gobbets' of Thucydides where, after reading his *Peloponnesian War*, we were given a tiny extract and asked to explain where it fitted in the story and what evidence it supplied. The training in, respectively, deductive and non-deductive reasoning was crucial.

I was found to have intellectual talents across both the humanities and science/mathematics. That is by no means an unmixed blessing, as it can impede the specialization essential to success in a modern academic career. To bring a mathematics/science perspective to the humanities or a humanistic perspective to mathematics risks being seen as an alien in both fields. On leaving school it posed an immediate problem of choice of direction. I had always expected to go into ancient history and so had chosen as a final school prize Liddell and Scott's massive *Greek–English Lexicon*, with a view to beginning Greek the next year at university. When I achieved first place in the Higher School Certificate in mathematics, I decided to change direction and enrol in pure mathematics – partly for its sound career prospects and partly because I doubted my ability to write prose well enough for humanities – although there was some concern as to whether I was sufficiently motivated by the mathematics itself.

An Arts degree at Sydney University required a choice of three first-year subjects. Mathematics was one, and I retained Ancient History in view of past interests. That left one more to find. I wondered about 'Philosophy'. I understood it was a very abstract subject, which was attractive, but I could not grasp what its subject-matter was. When the subject-matters of science, history, psychology and so on were added up, they seemed to exhaust the range of possible human knowledge. What was there left over for philosophy to be about? I enrolled to find out.

As it turned out, I did understand the philosophical point of view a little better than many of my peers because of my Catholic background. Catholicism has an elaborate theory and definite positions on the objectivity of truth and ethics. When younger I had accepted the Catholic faith from my mother and schoolteachers in a somewhat literal and conservative reading, but without much involvement. For example, I had no wish to be an altar boy. At the age of about 15, I experienced a kind of revelation, in retrospect as much ethical as religious, that meant I saw other persons as inherently valuable, psychologically interesting and imposing obligations on me, such as to help them if they needed it. That melded with the central aspects of the Gospel message, and I became more religious, though not much inclined to join religious groups or seminaries.

Reading some novels expanded my understanding of human possibilities (favourites included *Pride and Prejudice*, *The Brothers Karamazov* and, later, Anthony Powell's *Dance to the Music of Time*). However, I was not persuaded that literary criticism added value to simply reading literature.

My philosophical inclinations were also stimulated by a very unusual choice of set text, made in my final school year of Latin, Lucretius's *De Rerum Natura*. The straightforward, even adolescent, atomist atheism of the ancient poet, based on nothing but speculation and a wish to be free of gods, was interesting but also a warning on the attractions of swallowing a simplistic world picture.

Philosophy at Sydney University made quite an impact and I absorbed the methods of analytic philosophy as taught there, if not all the conclusions. The Philosophy Department included three very impressive philosophers, David Stove, David Armstrong and Graham Nerlich. Our introduction to Philosophy was Nerlich's course of lectures on free will and determinism, a topic that may not be elementary but certainly shows the relevance of philosophy – anyone who is not worried about the possibility that he/she is a puppet of past forces and brain chemistry does not have the philosophical spark. Nerlich's last lecture was surprising. He said that we first-year students would find it hard to appreciate, but worries about determinism were a mistake in modal logic, wrongly concluding from 'Necessarily, if conditions are X, I do Y' to 'If conditions are X, then necessarily I do Y'. This seemed to me a ludicrously wrong diagnosis. That sort of mistake would not happen in mathematics, I thought. Thereafter, I took a more critical attitude to anything said by my philosophy teachers.

Nevertheless, by the end of my first year, it seemed to me that the atheist materialism of my philosophy teachers had the best of the argument and that belief in Catholicism was not rational. I found that oppressive, unlike some of my contemporaries who found the release from 'priestcraft and superstition' liberating, especially the freedom it promised from 'conventional' moral constraints. But I was in for a shock. Second-year philosophy began with David Armstrong's lectures on Descartes's 'I think therefore I am'. Armstrong had published only four years earlier his very successful *A Materialist Theory of the Mind* and argued that Descartes' dualism of mind and body was entirely wrong. It seemed to me that Descartes was exactly right and had demonstrated that mind is an essentially different kind of thing from matter. Worldviews needed rethinking again. A universe with consciousness and objective ethics is not a materialist one. Some version of Catholicism, perhaps a little more liberal and without some of the unnecessary load of dogma, thereafter seemed the best overall theory of the universe, given the full range of evidence. A proper treatment of that argument is a project for the future.[1]

1973 was a big year for philosophy. As a third-year philosophy student, I had a front-row seat at the 'Sydney Philosophy Disturbances', where a strike with tents on the Quadrangle lawn demanded and eventually gained a feminism course. I later wrote the story in *Corrupting the Youth* inspired, like Thucydides, to write 'a history of my own time'.[2] As one of a conservative bent, I opposed the use of threats to achieve academic aims and I crossed some aggressive picket lines to attend Armstrong's lectures. But I attended the new feminism course to see what it was all about. It seemed to me a mystery worth inquiring into why so many powerful positions in society were occupied by men, but the course's Marxist talk about 'the ideology of bourgeois science' did not make progress on the question.

Also in 1973, I took a course with David Stove based on his then-new book *Probability and Hume's Inductive Scepticism*. Its main theme was that the relation of evidence to hypothesis in such cases as 'proof beyond reasonable doubt' in law or experimental evidence in science was a matter of strict logic (and hence necessary and accessible to intuition). I doubted that initially, but was soon convinced that it was not only true but an essential part of the answer to such irrationalist currents as postmodernism, then beginning to rear its head.[3] Logical probability or non-deductive logic was also found in the evidence for conjectures in pure mathematics, which confirms its logical nature since in pure mathematics there are no contingent principles such as the 'uniformity of nature'. My 1987 paper 'Non-deductive logic in mathematics' was to be my first major original work in philosophy.[4] Although it is rightly called a kind of probability, it does not follow that the relation of evidence to hypothesis can always be given a precise number – typically the evidence for a theory is intuitively strong or weak but there is no prospect of exact quantification.[5]

I was also encouraged by a fellow student, Geoffrey Harris, a Thomist philosopher trained at the Aquinas Academy in the city, to attend evening classes at that institution. Dr Austin Woodbury, who had studied in Rome in the 1920s with the hardline Thomist Réginald Garrigou-Lagrange, had run the Academy since 1945 as a philosophy school for lay people. By 1973, his last year of full-time teaching, he was somewhat past his prime. I found his personality overbearing and his approach formulaic and hard to understand, but the metaphysical questions were worthwhile and an interesting contrast to what I was learning at Sydney University. I resolved to read Aquinas to find out what it was really about. I made some progress – and I read some in Latin which was sometimes easier than in English – but it was still far from clear to me. So, I went back to Aristotle and finally understood; in particular, I understood and accepted the basic idea of the reality of properties of things as well as of the things themselves – for example, the law of gravitation connects mass, distance and gravitational force, all of which 'universals' are as real as the planets and moons that have them.

Surprisingly, at Sydney University, Armstrong was also developing an Aristotelian realism about universals. I took his 1974 honours course, later developed into his *Universals and Scientific Realism*, which came to be the leading treatment of the theme in analytic philosophy. One of his most revealing examples was the betweenness relations among colours. Given what red, orange and yellow are, it is absolutely necessary that orange is between red and yellow, and we have the intuition and reasoning powers to appreciate the necessity. While Aristotelian realism was a natural account of causality and laws of nature, mathematics was harder to fit into the story, as it appeared to be about Platonist entities like numbers and sets which were not universals in the real world – perhaps a gap in the market there.

The role of Dr Woodbury in the chain of events leading to my Aristotelian perspective, which formed the basis of much of what I did later, almost did not happen. In 1931, when at a Marist Fathers' institution in New Zealand, he had been called out to take a phone call; the Napier Earthquake struck and killed those he had just been with.[6]

The work of Aristotle that made the most impression on me was the *Posterior Analytics*. It laid out a vision of the ideal science as a structure derived from necessary and self-evident axioms by absolutely certain logical reasoning, which provided understanding by explaining why the truths of the science were as they were. It had proved impossible to make empirical sciences like physics and biology fit that model, as the basic principles were contingent and needed observational support. But the model was far from dead. Euclid's *Elements* followed the model well, while the more abstract disciplines like algebra, probability and ethics might still fit, and even in physics and biology more might be provable and understandable than at first was obvious.

Meanwhile, I continued to study pure mathematics with the intention of a PhD and academic career in that field. One aspect of the teaching of pure mathematics consistently annoyed me, its failure to explain what the various topics were about – including their historical motivations. Introducing a course on say 'modules' in 'definition, theorem, corollary, remark' style, with 'Example one: the set consisting of zero' provided no understanding. I did not demand usefulness in the sense of applications to physics or medicine but, in Aristotelian fashion, I sought a connection with the real world outside self-referring Platonic entities. One book was most helpful: Hermann Weyl's *Symmetry*, which described symmetry in art and biology and showed how the pure mathematics of group theory arose to study it.

I gained (just) an overseas postgraduate scholarship and completed a PhD in algebra (group representation theory), with one year at Durham University (where I visited the medieval cathedral every day and taught myself the basics of Western art history) and then at Warwick University. It was more or less successful[7] and I was fortunate to have an excellent supervisor, Roger Carter, but I was not sufficiently motivated to make it a life's work, and I did little further work in that area after I returned to Australia and took up a tutorship in the School of Mathematics at the University of New South Wales. I taught in that well-run School for almost forty years until my retirement, despite my research interests largely moving away from mathematics. My colleagues were grateful for my humanities expertise in 1997 when the University imposed a requirement that every degree programme should include a course on the professional issues and ethics of the discipline. No one in Mathematics could imagine what a course on 'Professional issues and ethics in mathematics' could possibly consist of, so I was left to create the world's first course on ethics in mathematics as I saw fit.[8] Actually, mathematics is a discipline with immense practical impact, for example in data analysis, and hence needs much more ethical scrutiny than it gets,[9] while ethics itself could, in some areas, make more use of mathematics, for example in accountancy which quantifies certain kinds of rights.[10]

While teaching mathematics at the University of New South Wales, it came to light that students could not understand what it meant to prove a result, and no one was telling them. The simplest matters of technique were not described, such as that to prove 'All even numbers have even squares' one should suppose 'x is even', explain what that means ('$x = 2k$, for some number k') and derive with some algebra 'x^2 is even'. After explaining this several times to the tutor in the neighbouring office, Albert Daoud, I agreed to his suggestion that we write a short textbook on it. *Introduction to Proofs in Mathematics* is an Aristotelian work in that it displays mathematical proof as deriving theorems from obvious truths by obvious logical steps, and also by seeing as the central logical notions the quantifiers 'all' and 'some' (as does Aristotle's *Prior Analytics*).[11]

The other thing revealed by mathematics teaching was that the skill of mathematical modelling, that is, of understanding and expressing in symbols the mathematical structure of some aspect of the real world, is quite a different skill from pure mathematics and is taught as rarely as proof. A very few classical mathematical models are well-known, such as the exponential growth model of population increase, but looking at a real problem and extracting its structure is not the same as imposing some pre-existing model. That is a process of Aristotelian cast too, as it finds abstract structure in the real world. It became clear to me through looking at examples such as Euler's study of the seven bridges of Königsberg that the provable necessities of mathematical structure were literally in the real world, not in some Platonic model of it: Euler proved that because of the topological structure of the bridges, one could not walk over the actual bridges once and once only.[12] One can actually see mathematical necessities in the real world. For example, a diagram of six crosses arranged as both two rows of three and three columns of two shows that $2 \times 3 = 3 \times 2$.[13]

Although I did not work further on algebra, I did 'keep my hand in' with some mathematical activities related to data analysis: in the 1990s in fuzzy logic and neural nets,[14] and in the 2000s in extreme risk (the effort to predict 'what hasn't happened yet', so there are no directly relevant data). These involve finding the mathematical structure in large datasets and require a combination of strictly statistical analysis with human judgement in the evaluation and combination of disparate evidence.[15]

I wrote about some of the major themes of mathematics, in a style that one referee called 'breezy synopticity'. A fault of mathematics degree programmes is that they do not include any course on mathematics – they have courses on algebra, calculus, probability and so on, but no course offering a conspectus of the subject. It is hard to see the wood for the trees, and the big picture is lost. It is significant, for example, that mathematics is situated in the wider field of the 'formal sciences' like operations research, statistics and theoretical computer science which, like mathematics, do not use the experimental methods of natural sciences but formal proof. Developed only since about 1940, they have greatly expanded the range of provable necessities.[16] Certain major themes of mathematics are 'well-known' but rarely written about in any generality. One is the discrete/continuous contrast, which divided ancient Greek mathematics into arithmetic and geometry and is still a major organizing theme, as well as having wider implications in, for example the atomism of chemistry versus the continuous nature (so far) of space-time.[17] Another is the local/global distinction, which is central to some of the core parts of mathematics such as algebraic topology. Some things that can be easily done locally, that is in small regions, cannot be extended to a whole structure; for example, in the case of Euler's bridges it is easy to walk across a few of the bridges once and once only, but those paths cannot be extended to a single path that goes over all seven bridges once and once only.[18] As the necessities of mathematics are absolute, they apply not only to our world (as the laws of physics do) but in all possible worlds, so mathematics attunes us to a deeper level of reality than empirical sciences.[19]

So, in the philosophy of mathematics, I looked for an Aristotelian realist philosophy that would start from applied mathematics and explain how mathematical necessities could be present in the physical world. Neither Platonism nor nominalism, the standard alternatives in the field, could do that. According to Aristotelianism, mathematics must be the study of some aspects of the real world, but it is not easy to say quite what aspects. My answer was laid out in *An Aristotelian Realist Philosophy of Mathematics: Mathematics as the Science of Quantity and Structure* (2014).[20] A few enthusiasts have joined me in the 'Sydney School' in the philosophy of mathematics.[21]

My Thomist study and historical interests had given me an appreciation of the contribution of the medieval and later scholastics to the development of modern thought, especially in disciplines where progress can be made by conceptual analysis.[22] They discovered that the human mind was not just a pure Cartesian ego but contained a great deal of complex implicit knowledge, which with effort could be brought to the surface of consciousness and become science.[23] One inspiring book on the topic was Carl Boyer's *History of the Calculus and Its Conceptual Development,* which showed how seventeenth-century calculus developed out of the scholastic analysis of continuous variation.

That combined with my interest in logical probability to provide the subject-matter of my first major – and still most successful – book, *The Science of Conjecture: Evidence and Probability Before Pascal.*[24] It examined the development of evidence evaluation up to the time Pascal and Fermat mathematized 'probability' in the sense of the theory of dice. The human mind is good at intuitive risk analysis, but making the principles conscious took centuries. The book proved a huge project – when halfway through and fully committed, I realized that I would have to read a considerable amount of medieval law in Latin, since the medieval law of evidence was the central topic, and that I would have to visit the nearest good university libraries to me to find the books. The libraries were in California.

I got to know David Stove better and on his death in 1994 he left me as his literary executor.[25] As a result, I researched more widely on Australian philosophy and wrote a polemical history of it up to the time of publication (2003), *Corrupting the Youth: A History of Philosophy in Australia.*[26] It emphasized the realist themes in Australian philosophy – stronger than in American and British philosophy – as well as personal stories and some of the aspects of farce (always stronger in philosophy than in, say, mathematics, as the line between deep insight and plausible nonsense is a thin one in philosophy). The title is not entirely ironic. A large part of philosophy does corrupt the youth.

While I am more interested in developing positive theories than in polemic, the 1990s were the years when academic life was especially threatened by the rising tide of crass philosophical errors loosely called 'postmodernism'. Postmodernism is essentially an attitude, an attitude of suspicion that aims to cut people off from the kind of contact with eternal verities that made possible my own education. I followed Stove in objecting to it and in trying to expose some of the philosophical mistakes[27] and by setting up a webpage on 'Australia's wackiest postmodernists'.[28] Needless to say, it had no effect. Mathematics itself stands as an unanswerable challenge to any relativist undermining of truth,[29] but humanities scholars show little interest in it.

In ethics, I became more convinced of its objective nature but less than happy about existing foundational efforts to explain why. The parallel between mathematics and ethics, in that both are abstract, non-experimental disciplines whose central truths are evident to intuition (very unlike the empirical sciences), evidently gave support to the objectivity of ethics since few doubted the solidity of mathematics (though some nominalist philosophies tried to deconstruct it as trivial).[30] On the foundations of ethics, Aristotle and Aquinas were not as satisfactory as on metaphysics. Like many other philosophers of ethics, they were misled by seeing ethics as fundamentally about what to do (actions, rights, virtues and so on). That is an aspect of ethics but not the deepest one. What is more important is what entities are inherently valuable or of worth. Once we know that, we can understand that actions supporting them are virtuous and actions harming them are wrong. I defended that view in *The Worth of Persons: The Foundation of Ethics* (2022).[31]

When young, I found debates on politics and ethics too confusing for a simple mathematician and philosopher to understand. People seemed to take positions very quickly and passionately, for reasons that could be appreciated, but without bothering to understand the reasons for thinking the opposite. The debates seemed too messy to make any progress with, unlike the clean lines of mathematics or metaphysics. Later I came to think that prospects were better – if we can put the opposing arguments in the abstract realm first, then we can make a judgement on the principles and their tradeoffs there and then work down to practical policy later.

For example, in economic theory, socialist theorists abusing 'rampant' capitalism and demanding fixes by state action faced Hayekian theorists believing that self-organization makes for the best outcome and that attempts to impose ethics on economics are both impossible and harmful. Those extremes do not exhaust the range of alternatives, nor do they describe the economic organization of the society we live in. I was impressed early with Pope Leo XIII's 1891 encyclical *Rerum Novarum*, for its argument that objective ethical requirements are as applicable to societal arrangements as they are to the actions of individuals. The history of economic regulation by the state and law (including international regulation[32]) since then has shown that it is possible to impose ethical requirements on economics, even though the problem of unintended consequences is serious. The economic system we live in is nothing like either 'rampant' capitalism or state socialism but is a kind of 'regulated capitalism' or 'market socialism', which works well though needs constant improving.[33]

Political argument too is given to clashes of free-floating one-sided ideological packages that do not gel with reality or engage with alternatives. Around 2000 I became disturbed by the mismatch between the well-known extreme levels of violence and poor health in remote Australian indigenous communities, and the obfuscatory Latinate discourse of city 'progressives' about 'reconciliation', 'colonialism', 'disadvantage' and 'self-determination'. The 250,000 who walked across the Sydney Harbour Bridge in 2000 'for reconciliation' had no wish to know about Skull, Walker and Ruben's 1997 article on 'Malnutrition and microcephaly in Australian Aboriginal children'.[34] I resolved to be on the lookout for better-grounded thinking. By chance, I came to know Jenness Warin, a remote-area nurse who could report on the appalling conditions in remote communities, and together we heard a talk by the Warlpiri leader Bess Price, which I believed finally gave a coherent view of the long-term levels of culturally approved violence in black communities and the need to restrain them. I supported Bess in her political career as a Minister in the Northern Territory Government, and later her daughter Senator Jacinta Price. I also supported the writing of Stephanie Jarrett's book *Liberating Aboriginal People from Violence* and set up an 'Australian Database of Indigenous Violence'.[35] Taking an abstract point of view made it possible to criticize collectivist ideological structures and see instead the individuals impacted by individual acts of violence and oppression.[36]

Catholicism is a religion that, more than most, is committed to intellectual work. It recognizes that philosophy and critical thinking are dangerous to faith but believes that the solution is more and better thinking, not the avoidance of thinking. Again, the basic idea of Pascal's Wager is a sound one: given the importance of the religion question and the real chance faith is true, an investment in thinking seriously and honestly[37] about it is rationally required.

That is especially so with the problem of evil, usually recognized as the most important intellectual problem facing belief in God. If God is good and omnipotent, how can he allow the horrendous evils we find in the world? It is evasive to say simply that it is 'a mystery' – does that mean that God has reasons, but we don't know them, or that there are no reasons? I came to understand that as mathematical necessities are absolute, they constrain God's omnipotence, so that omnipotence cannot accomplish things we wrongly think we can easily imagine, such as creating a better world by removing the evils of this one. That results in a view similar to Leibniz's theory that this is the best of all possible worlds – not that it is a good one, but that improvements to it are, to one who sees all mathematical necessities, impossible. A training in the necessities of mathematics gives some insight into a God's-eye view.[38] Some people might find the picture of an abstract and calculating God unfeeling and offensive to those who suffer evil, but the opposite is true, as it means that God is doing his best on our behalf.[39] My first wife Alex, who suffered from exceptionally bad health, came to a view something like that spontaneously.

After so long with necessities, it can be a relief to get back to history, which is full of contingent detail about what happened but could have been otherwise. The Catholic Church is a community of people as well as a theory, and a community needs to know its story as a tribute to those who came before, as role models and as warnings. I have edited the *Journal of the Australian Catholic*

Historical Society since 2011 and written extensively on the story of the Australian Catholic community.[40]

Of course, the journey was not as well planned as the map drawn at the end makes it appear. Many times, I pursued whatever intellectual lead seemed interesting at the time, like those dogs that lope down the street and stick their noses into whatever lies beside the track. But, in retrospect, the combination of my liking for the abstract and general with the gems left along the path by some of my teachers and earlier writers led to a result that is increasingly hard to come to in these days of postmodernist suspicion and academic over-specialization. The necessities of mathematics, logic and ethics are, at the bottom, the same kind of thing. Contemplating them gets us in touch with what really matters in the way that is right for humans – with real understanding.

Notes

(Author is James Franklin unless otherwise stated)

1 Preliminary sketch in *The Worth of Persons: The Foundation of Ethics* (New York: Encounter Books, 2022), epilogue; 'Emergentism as an option in the philosophy of religion: Between materialist atheism and pantheism', *Suri: Journal of the Philosophical Association of the Philippines* 7, no. 2 (2019): 1–22.

2 Also in 'The Sydney philosophy disturbances', *Quadrant* 43, no. 4 (Apr, 1999): 16–21.

3 'Resurrecting logical probability', *Erkenntnis* 55 (2001): 277–305.

4 'Non-deductive logic in mathematics', *British Journal for the Philosophy of Science* 38 (1987): 1–18; latest version in 'Bayesian perspectives on mathematical practice', in *Handbook of the History and Philosophy of Mathematical Practice,* ed. B. Sriraman (Cham: Springer, 2020), 2711–26.

5 'The objective Bayesian conceptualisation of proof and reference class problems', *Sydney Law Review* 33 (2011): 545–61; 'Case comment: quantification of the "proof beyond reasonable doubt" standard', *Law, Probability and Risk,* 5 (2006): 159–65.

6 Julie Thorpe, *Aquinas Academy, 1945–2015: A Very Personal Australian Story* (ATF Theology, Hindmarsh SA, 2016), 14.

7 'Homomorphisms between Verma modules in characteristic *p*', *Journal of Algebra* 112 (1988): 58–85.

8 A 'Professional issues and ethics in mathematics' course, *Gazette of the Australian Mathematical Society* 32 (2005): 98–100.

9 Dennis Müller, Maurice Chiodo and James Franklin, 'A Hippocratic Oath for mathematicians? Mapping the landscape of ethics in mathematics', *Science and Engineering Ethics* 28 (2022).

10 'Accountancy as computational casuistics', *Business and Professional Ethics Journal* 17, no. 4 (Winter, 1998): 31–37.
11 James Franklin and Albert Daoud, *Introduction to Proofs in Mathematics* (New York: Prentice Hall, 1988); *Proof in Mathematics: An Introduction* (Sydney: Kew Books, 2010).
12 'Mathematical necessity and reality', *Australasian Journal of Philosophy* 67 (1989): 286–94.
13 Catherine Legg and James Franklin, 'Perceiving necessity', *Pacific Philosophical Quarterly* 98 (2017), 320–43.
14 'How a neural net grows symbols', *Proceedings of the Seventh Australian Conference on Neural Networks*, Canberra, 1996: 91–96; 'Feature selection methods for solving the reference class problem', *Columbia Law Review* Sidebar, Mar 2010.
15 James Franklin, Scott A. Sisson, Mark A. Burgman and Jennifer K. Martin, 'Evaluating extreme risks in invasion ecology: learning from banking compliance', *Diversity and Distributions* 14 (2008): 581–91; 'Operational risk under Basel II: A model for extreme risk evaluation', *Banking and Financial Services Policy Report* 27, no. 10 (Oct 2008): 10–16.
16 James Franklin, 'The formal sciences discover the philosophers' stone', *Studies in History and Philosophy of Science* 25 (1994): 513–33.
17 'Discrete and continuous: a fundamental dichotomy in mathematics', *Journal of Humanistic Mathematics* 7 (2017): 355–78; 'Aristotle on species variation', *Philosophy* 61 (1986): 245–52; 'Achievements and fallacies in Hume's account of infinite divisibility', *Hume Studies* 20 (1994): 85–101; 'Arguments whose strength depends on continuous variation', *Informal Logic* 33 (2013): 33–56.
18 'Global and local', *Mathematical Intelligencer* 36(4) (Dec 2014): 4–9.
19 'Uninstantiated properties and semi-Platonist Aristotelianism', *Review of Metaphysics* 69, no. 1 (Sept 2015): 25–45.
20 *An Aristotelian Realist Philosophy of Mathematics: Mathematics as the Science of Quantity and Structure* (Basingstoke: Palgrave Macmillan, 2014); 'Mathematics as a science of non-abstract reality: Aristotelian realist philosophies of mathematics', *Foundations of Science* 27 (2022): 327–44.
21 The Sydney School: Mathematics, the Science of Structure, https://web.maths.unsw.edu.au/~jim/structmath.html
22 'Science by conceptual analysis: the genius of the late scholastics', *Studia Neoaristotelica* 9 (2012): 3–24; 'Mental furniture from the philosophers', *Et Cetera* 40 (1983): 177–91; 'Early modern mathematical principles and symmetry arguments', in *The Idea of Principles in Early Modern Thought*, ed. P. Anstey (New York: Routledge, 2017), ch 1.
23 Other examples in 'Diagrammatic reasoning and modelling in the imagination: the secret weapons of the Scientific Revolution', in *1543 and All That: Image and Word,*

Change and Continuity in the Proto-Scientific Revolution, ed. G. Freeland and A. Corones (Dordrecht: Kluwer, 2000): 53–115.

24 *The Science of Conjecture: Evidence and Probability Before Pascal* (Baltimore: Johns Hopkins University Press, 2001 and 2015) (given the lead review in the *Times Literary Supplement* on 11 June 2004 and praised by J.M. Coetzee and N.N. Taleb); summary in 'Prehistory of probability', in *Oxford Handbook of Probability and Philosophy*, ed. A. Hájek and C. Hitchcock (Oxford: Oxford University Press, 2016), 33–49.

25 Introduction to David Stove, *Cricket Versus Republicanism*, ed. J. Franklin & R. J. Stove (Sydney: Quakers Hill Press, 1995); 'Stove's anti-Darwinism', *Philosophy* 72 (1997): 133–36.

26 *Corrupting the Youth: A History of Philosophy in Australia* (Sydney: Macleay Press, 2003).

27 'The Sokal hoax', *Philosopher* 1, no. 4 (1996): 21–24; 'The Sokal hoax and postmodernist embarrassment', *Continuum* 14 (2000), 359–62; 'Thomas Kuhn's irrationalism', *New Criterion* 18 (10) (June 2000): 29–34; 'Stove's discovery of the worst argument in the world', *Philosophy* 77 (2002), 615–24; 'The postmodern calculus', *New Criterion* 29, no. 1 (Sept 2010): 75–80; 'Science matters in a post-modern world', in *Really Dangerous Ideas,* ed. G. Johns, (Ballan: Connor Court, 2013): 29–34.

28 'Australia's wackiest postmodernists', Mercatornet, June 2006, https://www.mercatornet.com/focus_on_postmodernism_australia_s_wackiest_postmodernists (accessed 10 May 2024).

29 'Last bastion of reason', *New Criterion* 18, no. 9 (May 2000): 74–78; 'Mathematics, core of the past and hope of the future', in *Reclaiming Education: Renewing Schools and Universities in Contemporary Western Society*, ed. C. Runcie and D. Brooks (Sydney: Edwin H. Lowe Publishing, 2018), 149–62.

30 'On the parallel between mathematics and morals', *Philosophy* 79 (2004): 97–119 (awarded 2005 Eureka Prize for Research in Ethics); '"Let no-one ignorant of geometry . . .": Mathematical parallels for understanding the objectivity of ethics', *Journal of Value Inquiry* 57 (2023): 365–84.

31 *The Worth of Persons: The Foundation of Ethics* (New York: Encounter Books, 2022).

32 'Risk-driven global compliance regimes in banking and accounting: the new Law Merchant', *Law, Probability and Risk* 4 (2005): 237–50; 'International compliance regimes: a public sector without restraints', *Australian Journal of Professional and Applied Ethics* 9, no. 2 (2007): 86–95.

33 'Regulated capitalism, market socialism', *Dissent* no. 5 (2001): 11–13; '*Caritas in Veritate*: Economic activity as personal encounter and the economy of gratuitousness', *Solidarity: The Journal of Catholic Social Thought and Secular Ethics* 1 (1) (2011).

34 Susan A. Skull, Alan C. Walker and Alan R. Ruben, 'Malnutrition and microcephaly in Australian Aboriginal children', *Medical Journal of Australia* 166 (1997): 412–14.
35 Australian Database of Indigenous Violence, archived at https://web.archive.org/web/20200818142612/http://indigenousviolence.org/dnn/
36 'The cultural roots of Aboriginal violence', *Quadrant* 52, no. 11 (Nov, 2008): 22–25; on collectivist errors, 'Global justice: an anti-collectivist and pro-causal ethic', *Solidarity: The Journal of Catholic Social Thought and Secular Ethics*, 2(1) (2012).
37 'Two caricatures, I: Pascal's Wager', *International Journal for Philosophy of Religion* 44 (1998): 109–14; 'Pascal's wager and the origins of decision theory: decision-making by real decision-makers', in *Classic Philosophical Arguments: Pascal's Wager*, ed. P. Bartha and L. Pasternack (Cambridge, Cambridge University Press, 2018), 27–44
38 'Two caricatures II: Leibniz's best world', *International Journal for Philosophy of Religion* 52 (2002): 45–56; 'The global-local distinction vindicates Leibniz's theodicy', *Theology and Science* 20 (4) (2022).
39 'Antitheodicy and the grading of theodicies by moral offensiveness', *Sophia* 59 (2020): 563–76.
40 *Catholic Values and Australian Realities* (Bacchus Marsh: Connor Court, 2006); *Catholic Thought and Catholic Action: Scenes from Australian Catholic Life* (Brisbane: Connor Court, 2023); *Arthur Calwell: Transforming Australia Through Immigration* (Brisbane: Connor Court, 2023); James Franklin, Gerry O. Nolan and Michael Gilchrist, *The Real Archbishop Mannix: From the Sources* (Ballarat: Connor Court, 2015).

2

Absolute necessities in the world and theories of necessity

Everyone's education equips them with vast knowledge of contingent matters of fact – that World War II ended in 1945, that the boiling point of water is 100°C, that knives go on the table to the right of forks. Everyone also knows many facts about the meaning and use of words, which lead to trivial necessary truths like 'All bachelors are unmarried' and 'What I have written I have written'. It is possible, in a modern education, to miss out on any serious exposure to more substantial necessary truths, facts about the way the world must be. Or at least, it is possible to learn of them as facts but accept them just on the basis of testimony, without understanding why they must be true. So let us start with several simple examples, chosen for variety, for immediate impact and for ease of understanding.

Example 1: Relations between numbers

Consider six crosses arranged in two rows of three crosses each, one row directly above the other.

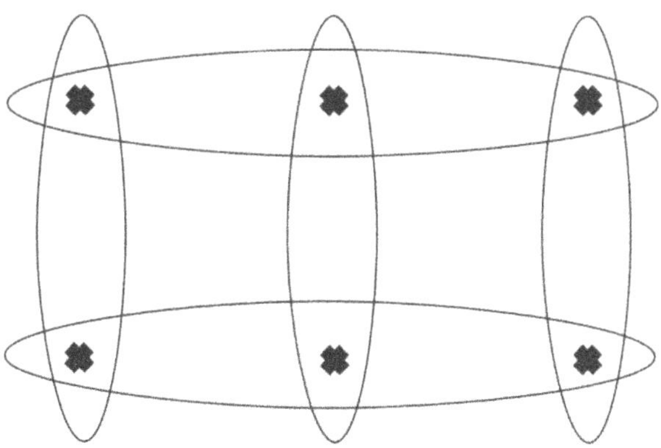

Figure 2.1 Why $2 \times 3 = 3 \times 2$.

The same six crosses form three columns of two each. Therefore 2 × 3 = 3 × 2. I not only notice that 2 × 3 is in fact equal to 3 × 2, I understand *why* 2 × 3 *must* equal 3 × 2.

Example 2: Network structure

In the eighteenth century, the seven bridges of Königsberg, in East Prussia, connected two islands and two riverbanks as shown in the diagram.

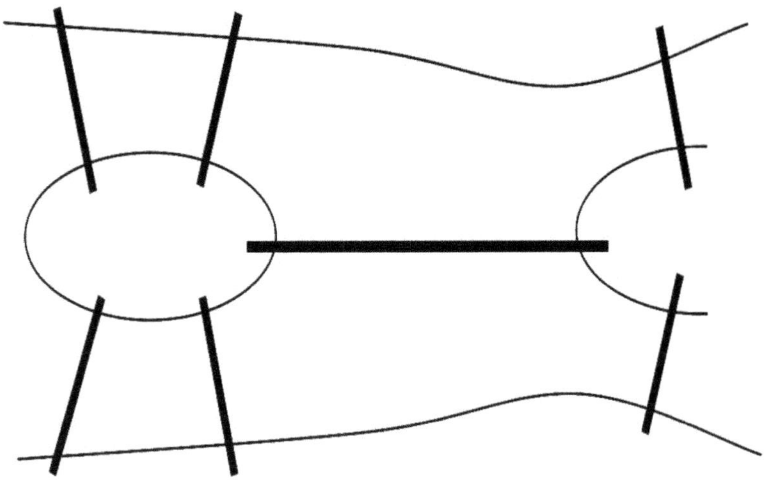

Figure 2.2 The bridges of Königsberg.

The citizens of Königsberg noticed that it seemed to be impossible to walk over all the bridges once, without walking over at least one of them twice. The mathematician Euler proved they were correct.[1] While new bridges could be built, with the existing bridges and land areas being as they are, it is absolutely impossible to walk over all the bridges once and once only. That is a necessity about the actual physical bridges and actual physical walking, not about some abstract model.

Example 3: Relations among properties

Many relations between properties are necessary. A typical example is the betweenness relations among colours. With red, yellow and orange being what they are, the fact that orange is between red and yellow is 'locked in': the colours could not be what they are without that betweenness relation holding. Another example is the transitivity of 'greater than' among quantities: if length A is longer than length B and length B is longer than length C, then it follows with absolute necessity that A is longer than C.

Example 4: Relation of evidence to hypothesis

Science cannot get under way without a commitment to the principle that true predictions by a theory confirm it. For example, if experiments agree with what a theory predicts, that is a positive for the theory. Polya describes as the 'fundamental inductive pattern':

> *q* is a (non-trivial) consequence of hypothesis *p*.
> *q* is found to be true.
> So, *p* is more likely to be true than before.

It is hard to begin reasoning about the world without a commitment to that principle. Imagine a tribe that did not believe in it and thought, instead, that *dis*agreement between theory and observation was a reason for believing the theory. They guess there are bison in the river field and go there to hunt them. They find none. So, they conclude they will probably find bison there tomorrow and the next day and they go there day after day with high hopes. You will need to imagine that tribe because you will not be meeting them. They are extinct.[2]

Experimental evidence does not imply with certainty that the theory that predicts it successfully is true. But the evidence-hypothesis relation in this and all cases is a necessary one: whether the evidence does or does not support the conclusion and, if so, how strongly, is a matter of necessity, not subject to differences of culture or opinion.

Example 5: (Non-trivial) logical necessity

Aristotle writes, 'the same attribute cannot at the same time belong and not belong to the same subject in the same respect'.[3] That is not a fact about language, it is how the world is and must be. What makes the argument:

> All As are Bs
> All Bs are Cs
> Therefore all As are Cs

valid is the (necessary) transitivity of the inclusion relation between sets. That is again a necessity about the world, not about language. Such 'prelogical' necessities are a precondition of the logic in the sense of properties of and relations between propositions, such as logical consequence.

Example 6: Ethical worth

What happens to humans can be a tragedy. What happens to rocks cannot. When we see pictures of atrocities, we understand that the victims were people like us, people who until their end possessed reason, apprehensions, personal unity, hopes, loves, and individuality as we do. That is what makes their death a tragedy and not, like the explosion of a lifeless galaxy, a firework. Gross metaphysical differences like that between humans and rocks imply, with strict necessity, gross moral differences. Thus, it is impossible to murder a rock, but possible to murder a human, and wrong to do so because it brings about the death of something that matters absolutely.

None of those necessities are dependent on language, or on conventions, or on the way culture has developed historically. They are part of the inherent structure of reality, prior to any human involvement or interests.

Unfortunately, in the standard philosophies of those fields – of mathematics, of probability, of logic, of ethics – those necessities are almost invisible. Subsequent chapters will rectify that by giving an account of each of the kinds of necessity of which those are examples. But first we survey the main lines of the philosophy of modality, as it has been pursued in the last few centuries and especially in recent decades. The main conclusion is that the philosophy of modality has made little sense of absolute necessities such as those in the examples and has barely made any attempt to do so. Standard philosophical thinking on the topic has kept such examples off the agenda, and for good reason, as it has few resources to deal with them. Nevertheless, considering this work gives some context to the examples, and shows how they compare to other kinds of necessity which have been better understood.

Types of necessity: trivial, nomic and absolute

While much has been written in philosophy about modality and its types, the field has become structured in such a way as to make barely visible the absolute necessities that are the subject of this book. They are missing even in the philosophy of modality (necessity and possibility) itself.[4]

We will briefly survey the types of necessity that are normally discussed, to clarify how those better-understood (or at least more frequently discussed) types of necessity differ from absolute ones.

Necessities come in three basic kinds: trivial ones created by concepts and language, 'nomic' ones expressed in the laws of nature, and absolute ones which are exceptionless ways the world must be.

The paradigm of trivial necessities is the linguistic truth 'All bachelors are unmarried'. It is true simply because of the meaning of the word 'bachelor' and does not express any extra- linguistic fact about the world. The same applies to the truth that for any proposition *p*, 'If *p* then *p*-and-*p*'. Repeating the assertion that *p* does not add anything contentful about the world (however contentful *p* itself may be); it is just how we use the word 'and', and the fact that, by linguistic convention, each occurrence of '*p*' denotes the same proposition.

The examples of absolute necessities given above are obviously not trivial in that sense. They are not generated by words or concepts and are contentful truths about their respective subject matters.

It is true that there could be trivial necessities that are complex and less than obvious. Perhaps some of the more complex truths of propositional logic are like that, and a large truth table is needed to check their triviality. Around 1900, Bertrand Russell's version of logicism hoped to establish that all mathematical truths were like that. That project is generally agreed to have failed, however. Technical problems included the inability to define set membership in logical terms, and the inability to show that the axiom of infinity (intuitively, that the whole numbers do not run out) was a logical truth.[5] But those technicalities were only a surface manifestation of the obvious fact that mathematics is about a non-logical subject matter, whether sets and numbers, or symmetry and ratio.

Ordinary language also recognizes an 'epistemic' use of necessity, 'it must be the case, given the accepted evidence'. 'You must have been a beautiful baby', given your present appearance; 'you must be Don Francisco's sister', given your dress and circumstances of arrival. Absolute necessities are obviously not of that nature,[6] although it is true that the logically necessary connections of evidence to conclusion are the basis of epistemic necessities and explain why modal language is used in such cases. An epistemic 'It must be that *p* (because we know *q*)' normally indicates a logically necessary connection (whether deductive or probabilistic, as in Example 4 above) between *q* (plus accepted background information) and *p*.

'Nomic' necessities, those encapsulated in the laws of nature, are certainly mysterious, but in a different way from absolute necessities. The most common way to think about the laws of nature, such as the inverse square law of gravitational attraction between any two bodies, is to see them as having a degree of necessity but one short of absolute. Such a law is more than a cosmic coincidence as it supports counterfactuals, such as 'If the earth had a second moon, its gravitational attraction to the earth would be exactly as described by the law'.[7] But it is usually thought that the laws could have been different and that they could be subject to miraculous exception[8] (as implied by the name 'laws', originally intended to suggest rules laid down by God to which he could choose to admit exceptions[9]). Neither of those, *prima facie*, are possible for the laws of arithmetic. Other views of laws of nature are possible, such as the Humean view, championed by David Lewis, that they are, in fact, just cosmic coincidences and reducible to mere patterns of facts,[10] and the opposite view of Aristotle, as interpreted by the medievals, that they do possess absolute necessity although that may be hard to know.[11] Here, however, since nomic necessity is not the subject under discussion, we will not inquire further.

The examples of absolute necessity given above are not nomic necessities. Changes in the laws of nature cannot change mathematical truths – that is why the philosophy of mathematics is so different from the philosophy of science. There is no apparent way that changes in the laws of nature or any changes to divine choices could affect logical relations, or the betweenness relation between colours. It is not out of the question that someone should attempt to devise possible worlds with different mathematical or logical laws, but the difficulties of describing them are evident and there is no substantial body of argument in favour of those possibilities.

It is significant, however, that even if individual laws of nature are not absolutely necessary, connections between them may be. Newton's mathematical derivation of Kepler's laws of planetary motion from the inverse square law of gravitation (in the absence of significant external forces) shows that the two apparently unconnected laws, even if contingent individually, have an absolutely necessary connection.

Not so mysterious are the *a posteriori* necessities that have come to be called 'Kripkean necessities'. Since 1970, a philosophical industry has grown up around a class of necessary truths discussed by Saul Kripke, and it has sometimes been maintained that they are paradigms of contentful necessities.[12] Classical examples are

Hesperus (the evening star) is Phosphorus (the morning star).

and

Water is H_2O.

In each case, there is only one thing: Hesperus is the same planet as Phosphorus, and water is the same substance as H_2O. But as a matter of epistemology, the different names for them arise from different ways of knowing them. The name 'Hesperus' refers to the bright heavenly body visible soon before sunset, while the name 'Phosphorus' refers to the bright heavenly body visible soon after dawn; 'water' refers to a natural stuff with a persistent complex of perceptible qualities including being liquid, colourless and hardly viscous, while 'H_2O' refers to the compound of two hydrogen atoms and one oxygen. It is facts about the world that determine, in each case, that the two modes of knowing converge on the same object: in one case, facts of celestial geometry, and in the other case, the nomic necessities of electron orbitals that determine that near-pure H_2O is (in our usual circumstances) liquid, colourless, hardly viscous and so on. Since those facts about the world are not obvious and can only be discovered by empirical investigation, the identities in question are rightly called *a posteriori*.

The absolute necessities under discussion are not of that kind either. They are not the result of different ways of knowing converging on an identity. Even in the case of $2 \times 3 = 3 \times 2$, where two different ways of dividing six objects do converge on an identity, the identity is immediately clear to vision and there is no need for empirical investigation to know it. In each case, the necessity is known by *a priori* understanding, not by empirical investigation.

Yet another sort of necessity frequently discussed arises from the distinction between the essential and accidental properties of a thing. It is said that being human is essential to Socrates, in that if he becomes non-human, he is no longer Socrates, whereas he can change colour while still staying Socrates. Thus, colour is an accidental property of him. By definition, 'an *essential property of an object* is a property that it must have, while an *accidental property of an object* is one that it happens to have but that it could lack'.[13] Whether that is correct or not, it is not relevant to the examples of absolute necessity above. The reason is that the essential-accidental distinction is premised on the Aristotelian-Linnean view of classification of substances: that the world is (approximately) uniquely divided into things (and stuffs) and those are uniquely divided into natural kinds.[14] However, the absolute necessities being considered are not in the category of substance at all, while in other categories the essential–accidental distinction does not apply. The truth that $2 \times 3 = 3 \times 2$ belongs in the category of quantity (in Aristotelian terms) and betweenness of colours in the category of quality. The absolute necessity of those truths has nothing to do with whether the quantities and qualities involved are essential or accidental to any substances. Even in the ethical case, which deals with qualities such as rationality that are often said to be essential to being human, it is not to the point whether they are, in fact, essential. It is the possession of those qualities themselves, whether essentially or otherwise, that gives humans moral standing and ethical worth.

It must be concluded, then, that the extensive discussion of necessities in recent philosophy, valuable as it is for the kinds of necessities it treats, has failed to focus on the sort of absolute necessities that are the subject of this book. The problem is evident in a prominent article by Kit Fine, 'The varieties of necessity'.[15] It proposes to divide necessities exhaustively into the logical, the metaphysical (meaning Kripkean and essentialist) the natural or nomic, and the normative (meaning about obligations). Absolute necessities are not mentioned. Some other account of those is needed.

To clear the path towards such an account, it is desirable to understand the theories of necessity (any kind of necessity) that one could have.

Theories of necessity: nominalist, Platonist and Aristotelian

The most prominent theories of necessity, nominalism and Platonism, have failed to give an adequate account of absolute necessities of the kind in the examples. As we will see in the chapters below on particular fields, there is in general a tripartite division in approaches to necessities: a nominalist, a Platonist and an Aristotelian realist one. Nominalism finds necessities only in words or concepts, not in the world. Platonism finds necessity in *a* world, but not the ordinary world we live in, only in a world of non-physical *abstracta* postulated precisely to house those necessities. Aristotelian realism, which will be defended across the board here, sees absolute necessities as existing in the physical world we live in (as well as in any other worlds there may be).

A nominalist approach to necessities arises from several sources. One is the hope that observed linguistic necessities like 'All bachelors are unmarried' will prove sufficient to cover all cases of necessity. That generally goes with a perception that ontological parsimony in metaphysics is a good in itself, as if a 'taste for desert landscapes' is in itself philosophically virtuous and should be defended against the 'bloated ontologies' favoured by philosophers with hidden agendas. That attitude, however, if a virtue at all, is a presumption of little weight. If there is no evidence for the existence of some sort of entity, such as witches, one should not believe in them, but if there is even a small amount of evidence for them, it must be considered and not dismissed.

The second source of nominalism is a logical-atomist picture of the world as consisting of loose and unconnected atoms, expressed in Hume's dictum, 'No necessary connection between distinct existences'. One thinks in the abstract as if, at least as an idealization, the world is the 'totality of facts', *a*'s being *F*, *b*'s being *G*, and so on, with any mix-and-match being possible, such as *a*'s being *G* and *b*'s being *F* instead. Hence there are no necessary connections between the 'distinct existences' *a*, *b*, . . . nor their properties *F*, *G*, . . .

The Humean dictum has been successfully attacked,[16] but it is worth seeing directly how the examples of necessities given above escape it. In the case of 2×3 = 3×2, the six crosses themselves are indeed distinct existences. But a necessary connection is asserted, not among them, but between the two rows of three and the three columns of two. Those are not distinct existences; on the contrary they wholly overlap: their constituting the same six total crosses is what makes the equation hold. And in the case of colours in the spectrum, such as red, orange and yellow, they are distinct existences, but their (necessary) placement in the spectrum ensures there are necessary connections among them, such as orange's being between red and yellow.

The clearest recent nominalist attempt to explain away real necessities is Amie Thomasson's *Norms and Necessity*. It is striking for the choice of examples taken as paradigms of the modal questions to be deconstructed as merely about language: *could* a person survive a brain transplant? What conditions *must* be met for a heavily restored painting to be identical to the original? *Can* a statue survive being squashed not the lump of clay it is made of?[17] *Prima facie*, those sorts of modalities may well depend on how we – we language users – decide to identify persons, paintings and statues. If that turns out to be true of those or Kripkean or essentialist necessities, that is no problem for the theory of this book. Those are not the sort of necessities being discussed here. Those necessities are not treated in nominalist theory.

At the opposite extreme to nominalism, Platonism accepts necessities, including absolute ones, but places them beyond the world we live in, in a Platonic realm of abstract entities which are non-physical and have no causal interaction with the physical world. Such 'full-blooded' Platonism is best-known in the philosophy of mathematics, where familiarity with the unchanging ideal world of numbers and sets has bred a conception of them as non-physical entities.[18] But it is also found in the 'Fregean Platonism' about propositions that is a default position in the philosophy of language, according to which logical relations like implication hold between abstract Senses or propositions shared by sentences in different languages which have the same meaning.[19]

Objections to Platonism have had over two millennia to mature. The most common one is epistemological – if Platonist entities are acausal and in another world, how could we know anything about them?[20] However from the present point of view, another objection is more fundamental – that as inhabitants of another realm, Platonist objects are not doing the job in this world that they were postulated to do. If triples of objects are paired off and found to be equinumerous, equality of their number has been established without reference to a number 3 in another world, and the existence of that number would be otiose. To put the point more precisely, what exactly is the relation between the things in this world (say triples) and the Platonist entity (say the abstract number 3) that enables the Platonist entity to perform its task in this world (of unifying triples, or explaining necessity or common meanings, or whatever it may be)?[21] What is it about triples that makes them and no other things have the correct relation to the abstract 3? Surely it is something inherent and this-worldly (such as being one of the equivalence classes of the pairing-off relation). And since the Platonist entities *are* in another, disconnected world, it makes sense to perform the thought experiment of imagining this world existing without the Platonic one. In that case the this-worldly properties of triples remain – including necessary ones such as every triple's containing three different pairs. The existence of Platonist entities is neither necessary nor useful for explaining this-worldly necessities.

Similar objections apply to a nominalist-Platonist hybrid that has become popular in discussions of modality in recent decades, David Lewis's realist theory of possible worlds. According to it, the Humean picture of disconnected atomic objects is true of the actual world, but necessity is outsourced to a system of all possible worlds, interpreted as necessarily existing Platonist entities. A proposition is necessarily true if it is true in all possible worlds (and that is all there is to being necessarily true). 'Modality is traded for quantification.'[22]

A natural reaction is expressed in the slogan, 'How do you know it's true in all possible worlds if you don't already know it's necessary?' but the issue is not epistemological, it is ontological. As Oderberg argues, 'any realist theory of possible worlds will be circular in its attempt to illuminate modality, for there must be some criterion of what counts as a *possible* world; there are by definition no *impossible* worlds. But then we have to have a prior conception of modality before we can use possible worlds to explain modality'.[23] Or to put the problem another way, what ensures that there are enough worlds (so that everything true in all worlds is in fact necessary) but not too many (so that some impossibilities slip in)? Surely an external notion of necessity is needed?[24]

Again, these problems are especially evident with the absolute necessities we have been considering. Surely the necessity of $2 \times 3 = 3 \times 2$, visually evident to us in this world, is what determines it to be true in all possible worlds (if there are any), and not vice versa.

It is perhaps Euler's Königsberg bridges example that poses the most obvious and most devastating counterexample to all non-Aristotelian theories of necessity, whether nominalist or Platonist. The inability to complete a path over all bridges just once is pre-linguistic and not affected by any changes in concepts or language. It is an inability of actual walkers in the actual Königsberg, so unaffected by any Platonic entities or what any counterparts are doing in any other possible worlds. Only a realist but non-Platonist theory – that is, an Aristotelian realist one – that sees necessity in the physical world can begin to make sense of it.

Aristotelian realism and its account of necessities

In order to develop an Aristotelian account of necessities, we first need to explain the Aristotelian realist perspective itself. Although Aristotelianism has a higher profile in, for example, the philosophy of science than in the philosophy of mathematics or of necessity, it is at odds with many current philosophical currents. Those who are trained in nominalism and Platonism can find it hard to grasp, despite its simplicity and its having appeared obvious in many previous centuries.

'*Orange is closer to red than to blue.*' That is a statement about *colours*, not about the particular things that have the colours – or if it is about the things, it is only about them *in respect of their colour*: orange things resemble red things but not blue things in respect of their colour. There is no way to avoid reference to the colours themselves.[25]

Colours, shapes, sizes, masses are the repeatables or 'universals' or 'types' that particulars or 'tokens' share. A certain shade of blue, for example, is something that can be found in many particulars – it is a 'one over many' in the classic phrase of the ancient Greek philosophers. On the other hand, a particular electron is a non-repeatable. It is an individual; another electron can resemble it (perhaps resemble it exactly except for position), but cannot literally be it.[26]

Science is about universals. There is perception of universals – indeed, it is universals that have causal power. We perceive an individual stone, but only *as* a certain shape, colour and weight, because it is those properties of it that confer on it the power to affect our senses. It is in virtue of being blue that a body reflects certain light and looks blue. Science gives us classification and understanding of the universals we perceive and finds the laws connecting them – physics deals with such properties as mass, length and electrical charge, biology deals with the properties special to living things, psychology with mental properties and their effects, mathematics with . . . we will examine that below.

Aristotelian realism about universals takes the straightforward view that the world contains both particulars and universals, and that the basic structure of the world is 'states of affairs' of a particular's having a universal, such as this page's being approximately rectangular.[27]

Science is also the arbiter of what universals there are. To know what universals there are, just as to know what particulars there are, one must investigate and accept the verdict of the best science (including inference as well as observation). Photosynthesis turned out to exist, phlogiston not. Thus, universals are not created by (or postulated to account for) the meanings of words, nor can one make up more of them by talking or thinking. On the other hand, language is part of nature, and it is not surprising if our common nouns, adjectives and prepositions name some approximation of the properties there are or seem to be (just as our proper names label individuals), or if the subject-predicate form of many basic sentences often mirrors the particular-property structure of reality. That is essential for understanding logic, as we will see in Chapter 6.

Aristotelian realism is committed to the reality of relations as well as properties. The relation being-taller-than is a repeatable and is accessible to observation in the same way as the property of being orange.[28] The visual system can make an immediate judgement of comparative tallness, even if its internal arrangements for doing so may be somewhat more complex than those for registering orange. Equally important is the reality of relations between universals themselves, such as betweenness among colours – if the colours are real, the relations between them are 'locked in' and also real. Western philosophical thought has had an ingrained tendency to ignore or downplay the reality of relations, from ancient views that attempted to regard relations as properties of the individual related terms to early modern ones that they were purely mental.[29] But a solid grasp of the reality of relations such as ratios and symmetry is essential for understanding how mathematics, especially, can directly apply to reality.

To say, however, that relations are real is not to say they are 'objects', like physical objects, that could have, for example, size or physical location. If one agreed that, in reality, London is west of Moscow and then asked 'But where is the relation "being west of"?', then one would have a true instance of Wittgenstein's dictum that 'Philosophical problems arise when language goes on holiday'. Relations, and universals generally, do not have the properties of physical objects.[30]

A large part of the general theory of universals concerns causality, dispositions and laws of nature, with all of these being related to (nomic) relations between universals.[31] A principal strength of Aristotelian realism is that it can give a natural account of the difference between a law of nature and a cosmic coincidence: a law that all As are Bs (unlike a coincidence that all As happen to be Bs) is the result of some real connection – the nature of which is to be discovered by science – between the real properties A and B. However, those nomic necessities are not the ones of interest here.

An Aristotelian theory of absolute necessity begins with the recognition that we have insight into instances of necessity that are not subject to exceptions in the way that nomic necessities (probably) are. Typically, they deal with a structured space of universals, like the range of determinates of the determinable colour. It is just the way colours are (and that is equally true whether they are at bottom mental or physical) that they come in a structured space with quasi-spatial structure, with a continuous range of colours between any two, and with some closer than others. (That is not to say the space of colours must be infinitely divisible, just that at the perceptible scale they vary continuously.) One can say that continuous variation and betweenness are just part of the nature of colours, but that need not be taken to imply an essentialist view of natures. Once we have come to recognize the colours red, orange and yellow, we understand their place in the colour spectrum and we need no empirical investigation to determine that orange is between red and yellow.

It is the same with numbers. Supposing we can give some Aristotelian account of how they exist as relations in the world (which will be done in the next chapter), we can understand how they must form a structured space with absolutely necessary relations between them – for example, that every quadruple consists of two pairs (in three possible different ways). Investigation is needed, but not empirical investigation. Similarly with ratios. The transitivity of 'greater-than' among ratios is locked in by the linear structure of the space of ratios: if ratio B is greater than A, and C is greater than B, then C is greater than A.

Those truths about the interrelations of ranges of universals of the same kind are absolutely necessary. They are not nomic. They are subject to understanding that does not require empirical investigation (except in the basic sense of perception allowing access to the universals in the first place).

Epistemology: how absolute necessities are known

While epistemology is not the main focus of this book, one of the main reasons for being interested in absolute necessities is that knowledge of them seems to take us in some sense beyond the perceptible world of the here and now. We are not taken into another, Platonic, world, but rather deeper into the one we live in – into the absolute necessities that underlie our familiar world.

That creates a problem with the knowledge of necessities. To know what is merely the case seems much easier than knowing that something must be the case and hints at mental powers beyond the ordinary. It may be suspected that one reason for the observed avoidance of absolute necessities by philosophers is a fear that the epistemology required by them may lead into dangerous territory. If the mind has a magical insight into why necessities must be so, who knows what challenges to naturalism may arise and what scarily immaterial properties may need to be attributed to the mind?

It would be more philosophically virtuous not to let a cringing prognostication impede present investigations. Let us follow where the reasons lead.

Nominalists are inclined to ask 'If we think of ourselves as trying to truly describe the modal facts or possible worlds (the presence of which could make the claims true), it is very difficult to see how we could have any (quasi-) perceptual or intellectual access to these modal facts or properties or worlds.'[32] It may be difficult to see *how* we could have such access, but it is not hard to (literally) see *that* we do – by looking at the $2 \times 3 = 3 \times 2$ diagram. 'How can we?' is a real question, not a rhetorical one.

Animals seem to have at best, only a vestigial 'must-detecting' (or possibility-detecting) faculty. They have trouble even with the simplest case of 'covering the bases' to catch a target that may come out of either of two holes but must come out of one of them.[33] Human capacities to understand necessities are significantly unlike anything in animals, and are not like artificial intelligence either – for all the amazing imitations of human cognitive output performed by AI, actual understanding is not found in AI and is very rarely aimed at. More will be said in the next chapter, since mathematical knowledge from pattern-recognition up to proof has been studied for millennia. Reflecting on that should make possible some progress on how to know absolute necessities.

Notes

1 Leonhard Euler, 'Solutio problematis ad geometriam situs pertinentis' (1735), trans. in *Graph Theory 1736–1936*, ed. Norman Biggs, E. Keith Lloyd and Robin J. Wilson (Oxford: Oxford University Press, 1976), 3–8.

2 James Franklin, *What Science Knows: And How It Knows It* (New York: Encounter, 2009), 9.

3 Aristotle, *Metaphysics* 1005b19–20.

4 See, e.g. Boris Kment, 'Varieties of Modality', *Stanford Encyclopedia of Philosophy*, 2012/2021; *The Routledge Handbook of Modality*, ed. Otávio Bueno and Scott A. Shalkowski (Abingdon: Routledge, 2021); *Logical Modalities from Aristotle to Carnap*, ed. Max Cresswell, Adriane Rini and Edwin Mares (Cambridge: Cambridge University Press, 2016).

5 Kevin C. Klement, 'Russell's Logicism', in Russell Wahl (ed.), *The Bloomsbury Companion to Bertrand Russell* (London: Bloomsbury, 2019), 151–78, especially section 6.

6 Martin Glazier, 'The Difference between Epistemic and Metaphysical Necessity', *Synthese* 198 (2021), 1409–24.

7 D. M. Armstrong, *What Is a Law of Nature?* (Cambridge: Cambridge University Press, 1983), 46–52.

8 Armstrong, *What Is a Law of Nature?*, 158–71.

9 Peter Harrison, 'Laws of God or Laws of Nature? Natural Order in the Early Modern Period', in Peter Harrison and Jon H. Roberts, eds, *Science Without God? Rethinking the History of Scientific Naturalism* (Oxford: Oxford University Press, 2019), 58–76.

10 Ned Hall, 'Humean Reductionism about Laws of Nature', in Barry Loewer and Jonathan Schaffer, eds, *A Companion to David Lewis* (Chichester: Wiley Blackwell, 2015), 262–77.

11 Edward Grant, 'The Condemnation of 1277, God's Absolute Power, and Physical Thought in the Late Middle Ages', *Viator* 10 (1979), 211–44, a modern revival in Chris Swoyer, 'The Nature of Natural Laws', *Australasian Journal of Philosophy* 60 (1982), 203–23.

12 Saul A. Kripke, *Naming and Necessity* (Cambridge, MA: Harvard University Press, 1980), Lecture III; overview in John P. Burgess, 'Kripke on modality', in *The Routledge Handbook of Modality*, ed. Otávio Bueno and Scott A. Shalkowski (London and New York: Routledge, 2021), ch 35.

13 Teresa Robertson Ishii and Philip Atkins, 'Essential vs. Accidental Properties', *Stanford Encyclopedia of Philosophy,* 2008/2020; survey in Penelope Mackie, 'Essentialism and Modality', in *The Routledge Handbook of Modality,* ed. Otávio Bueno and Scott A. Shalkowski (London and New York: Routledge, 2021), ch. 6; Bob Hale, *Necessary Beings: An Essay on Ontology, Modality, and the Relations Between Them* (Oxford: Oxford University Press, 2013), ch. 6; polar opposite views in David S. Oderberg, *Real Essentialism* (New York: Routledge, 2007); James Ladyman and Don Ross, *Every Thing Must Go: Metaphysics Naturalized* (Oxford: Oxford University Press, 2009).

14 James Franklin, 'Definition and Demonstration in the Category of Quantity and the Ancient Search for the Definition of Ratio', in Peter R. Anstey and David Bronstein, eds, *Definition and Essence from Aristotle to Kant* (New York: Routledge, 2025), 47–70.

15 Kit Fine, 'The Varieties of Necessity', in T. Gendler and J. Hawthorne, ed., *Conceivability and Possibility* (New York: Oxford University Press 2002), 253–81.

16 Jessica Wilson, 'What is Hume's Dictum, and Why Believe It?' *Philosophy and Phenomenological Research* 80 (2010), 595–637; Catherine Legg and James Franklin, 'Perceiving Necessity', *Pacific Philosophical Quarterly* 98 (2017), 320–43, section 2.

17 Amie L. Thomasson, *Norms and Necessity* (Oxford: Oxford University Press, 2020), 1.

18 James R. Brown, *Philosophy of Mathematics: A Contemporary Introduction to the World of Proofs and Pictures,* 2nd edn (New York, Routledge, 2008), ch 2; Øystein Linnebo, 'Platonism in the Philosophy of Mathematics', *Stanford Encyclopedia of Philosophy* (2009/2018).

19 Peter Hanks, *Propositional Content* (Oxford: Oxford University Press, 2015), 12–20.

20 Mark Balaguer, 'Platonism in Metaphysics', *Stanford Encyclopedia of Philosophy,* 2004/2016, section 5.

21 D. M. Armstrong, *Nominalism and Realism* (*Universals and Scientific Realism,* vol. 1) (Cambridge: Cambridge University Press, 1978), 66–9.

22 Scott A. Shalkowski, 'Logic and Absolute Necessity,' *Journal of Philosophy* 101 (2004), 55–82.

23 Oderberg, *Real Essentialism,* 2.

24 Scott A. Shalkowski, 'Modalism', in *The Routledge Handbook of Modality,* ed. Otávio Bueno and Scott A. Shalkowski (London and New York: Routledge, 2021), ch. 10

25 This and similar examples appeared in Arthur Pap, 'Nominalism, Empiricism and Universals: I', *Philosophical Quarterly* 9 (1959): 330–40.

26 Introductions to realist views on universals in J.P. Moreland, *Universals* (Chesham: Acumen, 2001), ch. 1; Chris Swoyer and Francesco Orilia, 'Properties', *Stanford Encyclopedia of Philosophy* (1999, revised 2011, archived), https://plato.stanford.edu/archives/win2011/entries/properties/.

27 D. M. Armstrong, *A World of States of Affairs* (Cambridge: Cambridge University Press, 1997).

28 D. M. Armstrong, *Universals and Scientific Realism* (Cambridge: Cambridge University Press, 1978), vol. 2 ch. 19.

29 Julius R. Weinberg, *Abstraction, Relation, Induction* (Madison: University of Wisconsin Press, 1965), part 2; Sarah E. Glenn, 'Relations and Reality: The Metaphysics of Parts and Wholes' (PhD diss., Boston College, Boston, 2000); David Odegard, 'Locke and the Unreality of Relations', *Theoria* 35 (1969): 147–52.

30 Discussion in Markku Keinänen, 'Lowe's Eliminativism about Relations', in *E.J. Lowe and Ontology*, ed. Miroslaw Szatkowski (New York: Routledge, 2022), 105–22 section 4.

31 D. M. Armstrong, *Truth and Truthmakers* (Cambridge: Cambridge University Press, 2004), ch. 10; D. M. Armstrong, *What is a Law of Nature?* (Cambridge: Cambridge University Press, 1983), part II; John Bigelow and Robert Pargetter, *Science and Necessity* (Cambridge: Cambridge University Press, 1990), chs 5–6.

32 Thomasson, *Norms and Necessity*, 17.

33 Jan M. Engelmann, Christoph J. Völter, Mariel K. Goddu, Josep Call, Esther Herrmann and Hannes Rakoczy, 'Invited Reply: Modal reasoning in non-human animals: possible ways forward', *Biology Letters* 19 (2024), 20230179.

Part Two

Mathematical Necessities

3

Aristotelian realist philosophy of mathematics

In looking for absolute necessities in the real world, mathematics is the natural place to start. Traditionally – since the time of Euclid – the certainty and objectivity of mathematics, its once-and-for-all establishment of rock-solid truths, has been Exhibit A in the museum of rationalism. Mathematics in the style of Euclid, with deductions from necessary axioms, has had a central place in education to train generations of students in solid deductions.

Mathematics has therefore stood as a challenge to many common philosophical positions. It is not just extreme sceptical views such as postmodernism that have a problem with it. So do all empiricist and naturalist views that hope for a fully 'scientific' explanation of reality and our knowledge of it. The problem is not so much that mathematics is true, but that its truths are absolutely necessary and that the human mind can establish those necessities and understand why they must be so. It is very difficult to explain how a physical brain could do that.

The problem is that the leading philosophies of mathematics in modern times, Platonism and nominalism, deflect attention from that, by representing the certainties of mathematics as confined to, respectively, a world of abstractions, or the world of human language and concepts. A new direction is needed in the philosophy of mathematics, which will reconnect mathematics with the world we live in, in particular its mathematical necessities. It needs to explain how mathematics can deliver that sort of necessary knowledge, while at the same time accounting for the extraordinary usefulness of mathematics in all kinds of scientific dealings with the real world – in a famous phrase, the 'unreasonable effectiveness of mathematics'.

Mathematics as a science of the world we live in

What is mathematics about? We know what biology is about; it is about living things. Or more exactly, the living aspects of living things – the motion of a cat thrown out of a window is a matter for physics, but its physiology is a topic for biology. Oceanography is about oceans; sociology is about human behaviour in the mass long-term; and so on. When all the sciences and their subject matters are laid out, is there any aspect of reality left over for mathematics to be about? That is the basic question in the philosophy of mathematics.

Philosophy of mathematics has been dominated by two classical but wrong answers. According to nominalism, it is not about anything, but merely a language or method. According to Platonism, it is about a world of "abstract objects" such as numbers and sets that exist in a non-physical realm. A later section will explain what is wrong with those views, but first, let us start with a physical world that has neither language nor *abstracta*.

Imagine the Earth before there were humans to think mathematics and write formulas. There were dinosaurs large and small, trees, eggs, volcanoes, flowing rivers, and winds. Were there, in that world, any properties of a mathematical nature (to speak as noncommittally as possible)? That is, were there, among the properties of the real things in that world (not some abstract world), some that we would recognize as mathematical?

There were many such properties. Symmetry, for one. Like most animals, the dinosaurs had approximate bilateral symmetry. The trees and volcanoes had an approximate circular symmetry with random elements – seen from above, they look much the same when rotated around their axis. The same goes for the eggs. But symmetry, whether exact or approximate, is a property that is not exactly physical. Non-physical things can have symmetry; arguments, for example, have symmetry if the last half repeats the first half in the opposite order. Symmetry is an uncontroversially mathematical property, and a major branch of pure mathematics – group theory – is devoted to classifying its kinds. When symmetry is realized in physical things, it is often very obvious to perception; if you have an asymmetrical face, don't go into politics, because it makes an immediate bad impression on TV. Symmetry, like other mathematical properties, can have causal powers, unlike *abstracta* as conceived by Platonists.

Another mathematical property, which like symmetry is realizable in many sorts of physical things, is ratio. The height of a large dinosaur stands in a certain ratio to the height of a small dinosaur. The ratio of their volumes is different – in fact, the ratio of their volumes is much greater than the ratio of their heights, which is what makes big dinosaurs ungainly and small ones sprightly. A given ratio is something that can be the relation between two heights, or two volumes, or two-time intervals; a ratio is just what those relations between different kinds of physical entities share and is thus a more mathematical property than the physical lengths, volumes and so on. Ratio is what we measure when we determine how a length (or volume, or time, etc) relates to an arbitrarily chosen unit. It is one of the basic kinds of number. As Isaac Newton put it in his uniquely magisterial language: 'By Number we understand not so much a Multitude of Unities, as the abstracted Ratio of any Quantity, to another Quantity of the same kind, which we take for Unity'.[1] But despite its abstract and general nature, ratio is easily perceivable, as in this classic diagram of body types where the different ratios of height to width are very perceptually salient:

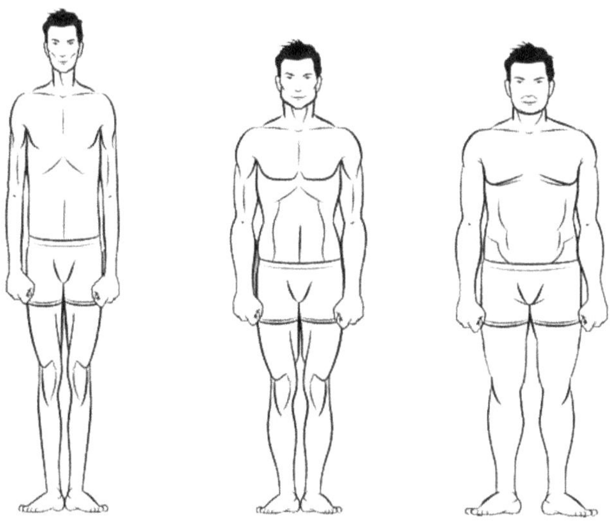

Figure 3.1 Direct perception of ratio.[2]

Any digression into applied mathematics – rarely undertaken by philosophers of mathematics, who prefer the familiar ground of numbers and logic – will turn up, for the alert observer, many other quantitative and structural properties[3] that are not themselves physical but can be realized in the physical world (and in any other worlds there might be): flows, order relations, continuity and discreteness, alternation, linearity, feedback, network topology, and many others.

The philosophy of mathematics that emphasizes how mathematical properties crop up in the actual world is called Aristotelian realism. It is based on Aristotle's view, opposed to that of his teacher Plato, that the properties of things are real and in the things themselves, not in another world of *abstracta*. A version of it, holding that mathematics was the 'science of quantity', was actually the leading philosophy of mathematics up to the time of Newton, but the idea has been largely off the agenda since then.

Two clarifications are needed. To say that mathematical properties can be realized in the physical world is not to say that they can be realized *only* in the physical world. Palindromes have symmetry and, as noted above, arguments can too. Nor is it to say that any given mathematical property *must* be instantiated at all. Just as there could be an uninstantiated shade of blue – that is for the world to decide, not the nature of colours – so there could be uninstantiated mathematical properties, such as higher infinities.[4]

Mathematical necessities in physical reality

An essential theme of the Aristotelian viewpoint is that the truths of mathematics, being about universals and their relations, should be both necessary and about reality. That thesis stands in opposition to the common view expressed in Einstein's classic dictum, 'As far as the propositions of mathematics refer to reality, they are not certain; and as far as they are certain, they do not refer to reality.'[5] Einstein made it clear that by 'certain' he meant 'necessary', and philosophers of recent times have mostly agreed with him that there cannot be mathematical truths that are at once necessary and about reality.

Mathematics provides, however, many *prima facie* cases of necessities that are directly about reality. Perhaps the most classic and clearest example is Euler's work on the bridges of Königsberg, mentioned in the previous chapter. The impossibility of walking over all the bridges once and once only is provable and absolute, and it is about the actual system of bridges, not about some idealized model. Some further examples will be given in the next chapter – along with a simple proof of the bridges of Königsberg result.

What are numbers?

While an Aristotelian realist philosophy of mathematics does not regard whole numbers as a specially important part of mathematics, compared to ratio, symmetry, network topology and so on, it obviously should provide an account of them. That account should explain where they can be found in the physical world and what use they are in dealing with that world. It should explain how the truths about numbers, which are generally accepted to be necessary, retain their necessity when found in the real world.

We adopt the theory of Glenn Kessler, which builds on a gnomic remark of Aristotle that "'The one' means the measure of some plurality, and 'number' means a measured plurality . . . the measure must always be some identical thing predicable of all the things it measures, e.g. if the things are horses, the measure is 'horse' . . ."[6] He means that a number is a *relation* – a relation of some part of the world, such as some horse-mass, to a 'unit-making' universal, such as 'being a horse', that structures the mass into so many individual horses. Those universals are the ones whose names are typically 'count nouns', like 'horse' and 'apple', with singulars and plurals, as opposed to 'stuff' or abstract nouns like 'water' and 'justice', but the question is not a linguistic one – apple mass is structured into individual apples prior to language.

Frege rightly observes that the *Iliad* is one poem, or 24 books, or a large number of verses and rightly concludes that number is not a property of the *Iliad*. But then he wrongly concludes that therefore a number is not in the world at all,[7] instead of reaching the obvious conclusion that it is a *relation* in the world: the *Iliad* has one numerical relation to being a poem, another to being a book, and a third to being a verse.[8] Thus numbers are not properties of parts of the world simply, but are relations between parts of the world and the unit-making properties that structure them.[9] The numerical relation between certain rare clover leaves and 'being a clover leaflet' is the same as that between the sum of horsemen of the Apocalypse and 'being a horseman of the Apocalypse'; namely 4. Counting proceeds when we choose a unit, say being a shoe, and determine the repetitions of that unit in the heap to be counted; choosing a different unit, say being a pair of shoes, will produce a different relation to the unit from the same heap. There is nothing subjective about the relation: given a heap and a unit-making universal, there is just one objective numerical relation, how many of that unit-making universal are found in that heap.

We can then see why the necessities that are proved in number theory apply directly, and as necessities, in the real world. Numbers just are relations in the real world, so any necessary truths about them apply wherever numbers are found. For example, the fact of 'pure mathematics' that $4 = 2 + 2$ is the same fact as that any quadruple is composed of two pairs.

Nominalist and Platonist errors

What has just been said is very far from the usual diet in the philosophy of mathematics. The field is dominated by two alternative views, nominalism and Platonism, which are incompatible with Aristotelian realism (as well as with each other). While something was said in the introduction about nominalism and Platonism in general, the problems with them come into sharper focus when we consider them as applied to mathematics – a field where, indeed, much of the debate has taken place.

Nominalists hold that mathematics is just a language. On this view, it is just a way of talking about other things, or a collection of logical trivialities, or what follows from arbitrarily chosen axioms, or a formal manipulation of symbols according to rules. On any of those views, it is not really about anything. Those whose encounter with mathematics at school was less than happy ('Minus times minus equals plus/The reason for this we need not discuss') might feel some sympathy with the nominalist picture. It is also a view that appeals to physicists and engineers who regard serious propositions about reality as their business. They look on tables of Laplace transforms and other such mathematical paraphernalia as, in the words of the German philosopher Carl Hempel, 'theoretical juice extractors':[10] useful for revealing the implications of meaty physical propositions, but not contentful in themselves.

Nominalism may have a certain down-to-earth appeal, but further reflection suggests that it cannot be right. Although manipulation of symbols is useful as a technique, we also have a strong sense that mathematics makes objective discoveries about a terrain that is in some sense 'out there'. That is true of applied examples like the Königsberg bridges but perhaps better appreciated in more 'pure' examples. Take the subtleties of the distribution of primes. Some numbers are prime, some not. A dozen eggs can be arranged in cartons of 6×2 or 3×4, but eggs are not sold in lots of 11 or 13 because there is no neat way of organizing 11 or 13 of them into an egg carton: 11 and 13, unlike 12, are prime, and primes cannot be formed by multiplying two smaller numbers. The idea of primeness is easy to grasp. But its implications are complex.

It turns out that the way in which the primes are distributed among numbers involves a subtle interplay of pattern and irregularity. On the small scale, the latter is most evident: there are long stretches without any primes at all – indefinitely long stretches, in fact. At the same time, it is widely believed (though not yet proved) that there are infinitely many 'prime pairs', that is, pairs of numbers only two apart that are both prime, such as 41 and 43.

When we turn to the large scale, the impression of disorder fades and a pattern starts to emerge after all. Primes become gradually less dense as one counts up: the density of primes around a large number is inversely proportional to its order of magnitude. Thus, the density of primes around a trillion (10^{12}), for example, is about half what it is around a million (10^6). More exact information on the intricacies of the distribution of primes is contained in the Riemann Hypothesis, currently the most famous unproved conjecture of mathematics.

That is typical of the results of pure mathematics, from simple school facts such as the divisibility of numbers by 9 if the sum of their digits is divisible by 9, up to the higher reaches of abstract algebra. It is impossible to escape the conclusion that pure mathematics reveals to us the topography of a region whose truths pre-existed our investigations and even our language. Nominalist approaches fail to account for that.

Inspired by that thought, Platonism proposes a philosophy of mathematics opposite to nominalism. It says that mathematics is about a realm of non-physical objects such as numbers and sets, *abstracta* that exist in a mysterious realm of forms beyond space and time. If that sounds far-fetched from a philosophical perspective, it is true that pure mathematicians certainly speak and often think that way about their subject. Platonism also fits well with the apparent success of mathematical proof, which seems to demonstrate how things must be in all possible worlds, irrespective of what the laws of nature might be in any particular world. The proof that the square root of 2 is an irrational number does not rely on any observationally established laws or experimental measurements. It shows how things must be, suggesting that the square root of 2 is an entity beyond our changeable world of space and time.

Still, despite its clean lines and long history, Platonism cannot be right either. Since the time of Plato himself, nominalists have been urging very convincing objections. They argue that abstract objects are ruled out by naturalism and are not needed in any other area of science, so mathematics would be anomalous in requiring their existence. Again, if *abstracta* float somewhere outside our own universe of space and time, it is hard to imagine how we can see them or have any other perceptual contact with them. So how do we know they exist? Some contemporary Platonists claim that we infer them, much as we infer the existence of atoms to explain the results of chemistry experiments.[11] But that seems not to be how we know about numbers. Five-year-olds learning to count do not perform sophisticated inferences about abstractions; their contact with the numerical aspect of reality is somehow more perceptual and direct. Even animals can count, up to a point.

In any case, the problem with Platonism is not so much about knowledge as about its view of mathematical entities. Surely when we measure, or calculate, or model the weather mathematically, we are dealing with mathematical properties of real things in this world, such as their quantities? Such properties are not *abstracta*: like colours, they have causal powers that result in our seeing them. The visual system easily detects such properties as the ratio of your height to mine (if we stand next to each other). There is no room for *abstracta* in other worlds to enter the story, even if they did exist.

Nominalists and Platonists have fought each other to a standstill, each convincingly revealing the fatal flaws in their opponents' views, each unable to establish their own position. They have failed to account for the simplest facts about how mathematics describes the world we live in – nominalism by reducing mathematics to trivialities, and Platonism by divorcing it from the real world of which mathematical truths form a necessary skeleton. Aristotelian realism is a new beginning. It connects the philosophy of mathematics back to the applications that have always been the fertile ground from which mathematics grows. It has a message both for philosophy and for mathematics and its teaching: don't get blinded by shuffling symbols, don't disappear into a realm of abstractions, just keep an eye fixed on the mathematical structure of the real world.

In addition, Aristotelians believe that Platonists and nominalists make a common error: of regarding everything that exists as a particular (whether abstract or concrete). As explained in the introduction, Aristotelians argue that science typically deals in universals – for example the relations between mass, distance and force described in Newton's law of gravity – and that mathematics should follow suit by, for example, studying properties like symmetry and continuity and what progressions have in common. Universals are not "things" (objects, particulars), whether abstract or concrete. They do not come into and go out of existence (though particulars can come to and cease to have universals). Thus, the truths about them, such as the transitivity of greater-than (if A is greater than B and B is greater than C, then A is greater than C) can be eternal and necessary. Aristotelianism is thus not a simple empiricism, despite claiming that mathematics is directly about reality.

Knowing mathematics

Because Aristotelian realism insists on the realizability of mathematical properties in the world, it can give a straightforward account of how the most basic mathematical facts are known: by perception, the same as other simple facts. Ratios of heights are visible (to a degree of approximation, of course). Infants and animals demonstrably do have the ability to recognize patterns and estimate numbers, shapes, and symmetries.[12]

Our developed human intellectual abilities add two things to those simple perceptions. The first is visualization, which allows us to understand necessary relations between simple mathematical facts. That was what allowed us to see that $2 \times 3 = 3 \times 2$ (see Fig 3.2).

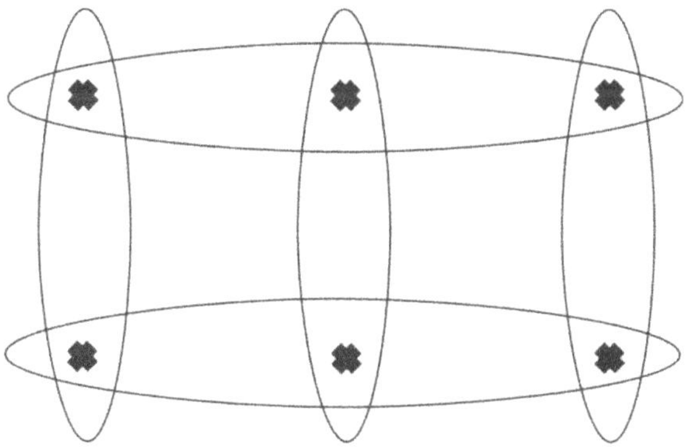

Figure 3.2 Why $2 \times 3 = 3 \times 2$.

The same six crosses form two rows of three each and three columns of two each. Therefore, we can see (literally) *why* 2×3 must equal 3×2.[13] So the Platonists were right to call attention to the ability of the human mind to grasp mathematical necessities; they just failed to notice that those necessities are often realized in this world. The second intellectual ability by which the human mind extends the results of perception is proof. As described in the next chapter, mathematical proofs chain together a series of insights, individually similar to '$2\times3 = 3\times2$', to demonstrate necessities that cannot be understood at a glance, such as how the density of primes tails off for large numbers.[14]

Aristotelian realism stands in a difficult relationship with naturalism, the project of showing that all the world and human knowledge can be explained in terms of physics, biology, and neuroscience. If mathematical properties are realized in the physical world and capable of being perceived, then mathematics can seem no more inexplicable than colour perception, which surely can be explained in naturalist terms. On the other hand, Aristotelians agree with Platonists that the mathematical grasp of necessities is mysterious. What is necessary is true in all possible worlds, but how can perception see into other possible worlds? The scholastics, the Catholic Aristotelian philosophers of the Middle Ages, were so impressed with the mind's grasp of necessary truths as to conclude that the intellect was immaterial and immortal. If today's naturalists do not wish to agree with that, they face a challenge to do better. 'Don't tell me, show me': build an artificial intelligence system that imitates genuine mathematical insight. Despite the amazing abilities of AI in other areas, including even discovering mathematical proofs, there seem to be no promising plans on the drawing board to achieve genuine understanding.

Conclusion

Mathematics really does have all the advantages attributed to it by a naïve view: absolute necessity, provable certainty leading to insight on why things must be so, and direct applicability to physical and other reality. That needs to be taken into account not only by the philosophy of mathematics, but also by wider philosophy. It would be unwise to accept any general philosophy which implied that such a combination of virtues was impossible.

Notes

1 Isaac Newton, *Arithmetica Universalis* (London, 1728), 2.

2 Source: https://en.wikipedia.org/wiki/Somatotype_and_constitutional_psychology#/media/File:Bodytypes.jpg

3 Exact definitions of 'quantitative' and 'structural' are given in James Franklin, *An Aristotelian Realist Philosophy of Mathematics: Mathematics as the Science of Quantity and Structure* (Basingstoke: Palgrave Macmillan, 2014), chs 3 and 4.

4 James Franklin, 'Uninstantiated Properties and Semi-Platonist Aristotelianism', *Review of Metaphysics* 69 (2015): 25–45; *An Aristotelian Realist Philosophy of Mathematics*, ch. 2.

5 Albert Einstein, *Ideas and Opinions* (New York: Random House, 1954), 233.

6 Aristotle, *Metaphysics* bk 14 ch 1, 1088a4–11.

7 Gottlob Frege, *The Foundations of Arithmetic*, §22 (trans. J. L. Austin, 2nd revised edn, Oxford: Blackwell, 1980), 28.

8 Andrew D. Irvine, 'Frege on Number Properties', *Studia Logica* 96 (2010): 239–60.

9 Glenn Kessler, 'Frege, Mill and the Foundations of Arithmetic', *Journal of Philosophy* 77 (1980): 65–79; developed in Peter Forrest and D. M. Armstrong, 'The Nature of Number', *Philosophical Papers* 16 (1987): 165–86.

10 Carl G. Hempel, 'On the Nature of Mathematical Truth', *American Mathematical Monthly* 52, no. 10 (1945): 543–56.
11 Mark Colyvan, *The Indispensability of Mathematics* (Oxford: Oxford University Press, 2001).
12 For the numerical sense the classic exposition in Stanislas Dehaene, *The Number Sense: How the Mind Creates Mathematics* (New York: Oxford University Press, 1997); at a lower level, M. Giurfa, B. Eichmann and R. Menzel, 'Symmetry Perception in an Insect,' *Nature* 382, no. 6590 (1996): 458–61.
13 Further in Catherine Legg and James Franklin, 'Perceiving necessity', *Pacific Philosophical Quarterly* 98 (2017): 320–43; Franklin, *An Aristotelian Realist Philosophy of Mathematics*, ch. 11.
14 James Franklin, *An Aristotelian Realist Philosophy of Mathematics*, ch. 12; An Aristotelian perspective on proof in James Franklin and Albert Daoud, *Proof in Mathematics: An Introduction* (Sydney: Kew Books, 2010).

4

Proof in mathematics and beyond

Euclid's *Elements* have played a leading role in education for over 2,000 years. The reason is that they provided training in the strictly logical derivation of complex and unobvious geometrical theorems, like Pythagoras's Theorem, from simple and evident truths like the definition of a circle. Someone who follows even the first few proofs in Euclid understands the logical structure of a large body of truths, that is, the absolutely necessary connections between them.

Before looking at how that worked and how mathematical proofs are found beyond pure mathematics, in computer science and applied mathematics, we will exhibit two very easy proofs in areas of mathematics that are simpler than geometry, namely number theory and combinatorics. They will make it clear how absolute necessities can be established, free of any empirical or otherwise dubious assumptions.

Two simple proofs

Theorem 1: The square of any even number is even (equivalently, 'the square of every even number is even'; 'the squares of all even numbers are even'; 'if a number is even then its square is even').

It is easy to check many examples of the square of an even number being even: $2^2 = 4$, $12^2 = 144$, $1230796^2 = 1514858793616$ . . . But that provides no guarantee that all the infinitely many even numbers will have even squares, nor any understanding of why it should be so. In empirical science, one must be content with evidence from many examples, but mathematics is different.

To enable the proof to be general – applicable to all even numbers whatsoever – we need to start with an arbitrary even number and use only the fact that it is even. So, the outline of the proof must look like this:

Proof: Let x be even

$$\Downarrow$$

So x^2 is even

To make progress, we need some definition or analysis of 'even', so that known methods can be applied to make the transition from truths about x to truths about x^2. The standard analysis is that a number is even if it is double some other (whole) number. That is a non-trivial step and needs some insight, if one's initial vague conception of even numbers is just as of every second number after 2, or a number that can be cut evenly into two parts. So, the outline of the proof must be filled in the second and second-last lines as follows:

Proof: Let x be even

So $x = 2 \times k$ for some integer k (definition of 'even')

$$\Downarrow$$

So $x^2 = 2 \times ?$

So x^2 is even

At that stage it is clear how to proceed to fill in the gap: one should square both sides (if numbers are equal, their squares must be equal). A little standard algebraic manipulation (using only facts about the irrelevance of order in multiplication) gives the complete proof:

Prove that the square of any even number is even

Proof: Let x be even

So $x = 2 \times k$ for some integer k (definition of 'even')

So $x^2 = (2 \times k)^2$

$= (2 \times k) \times (2 \times k)$

$= 2 \times (k \times 2 \times k)$

So x^2 is even (definition of 'even')

The proof is simple enough to be visualizable (in a particular example, of course, not in general, although our visualization ability mysteriously permits a kind of implicit generality).

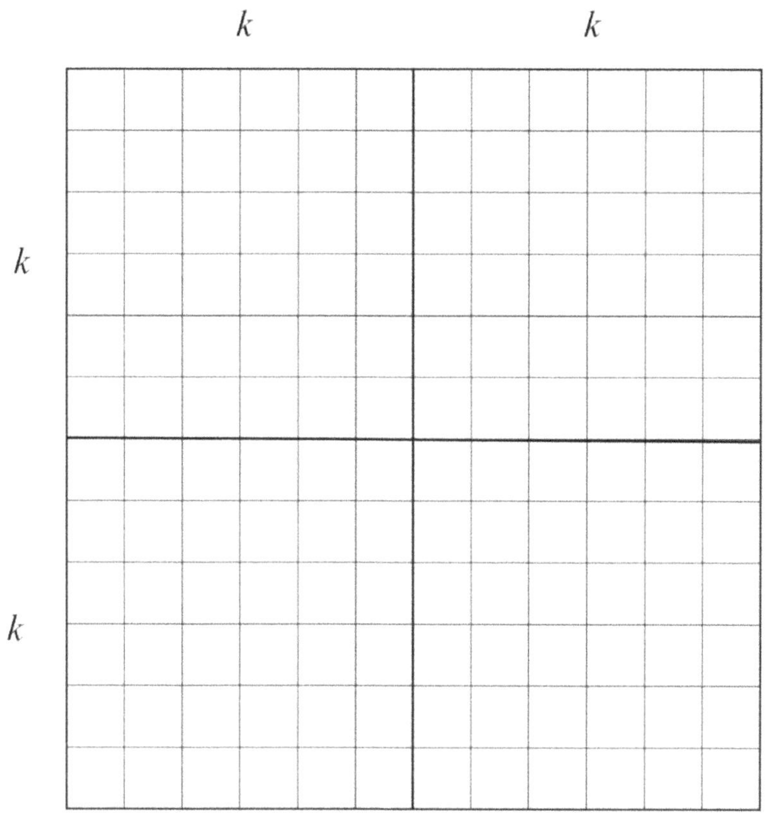

Figure 4.1 The square of any even number is even.

If the side is even, that is, consisting of two equal parts, so is the square. In fact, it is clear from the diagram that the square of any even number is divisible by 4 – as can be confirmed by the algebraic proof.

The example confirms another discovery of Aristotle's: that the crucial things in the logic of real reasoning are the two basic quantifiers, 'all' and 'some'.

Next, we take a 'bare hands' proof from combinatorics, an area of mathematics particularly uncluttered with possibly doubtful assumptions (for example, about infinity).

Theorem 2: Consider six points, with each pair joined by a line. The lines are all coloured, in one of two colours (represented by dotted and undotted lines in the figure). Then there must exist a triangle of one colour (that is, three points such that all three of the lines joining them have the same colour).

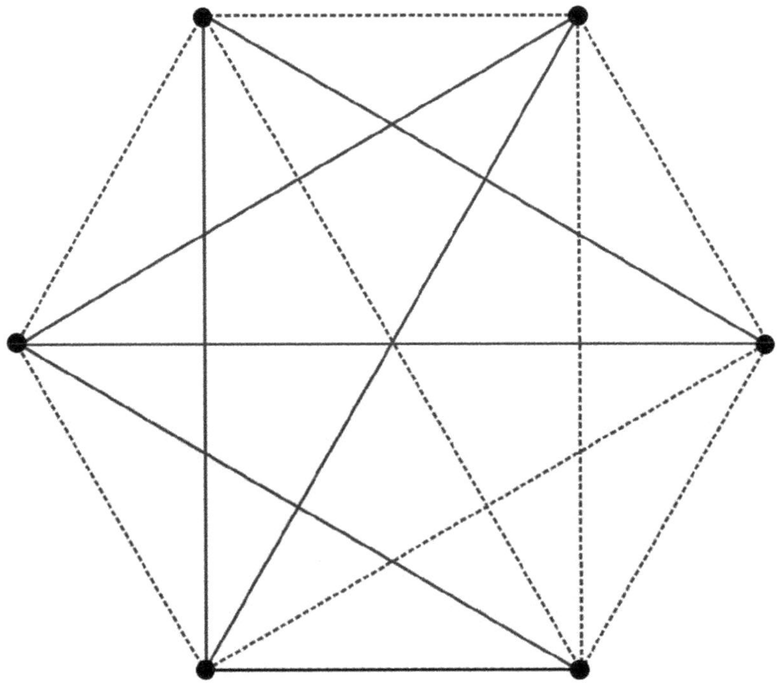

Figure 4.2 Combinatorics with six points.

Note that this result is not suitable for a one-glance visualization 'proof', similar to the one above or the one in the last chapter that enables us to understand why $2 \times 3 = 3 \times 2$. The reason is the complexity of the possible configurations. We might just be able to visualize a single configuration, as in the diagram. There is no hope of visualizing all of them, and seeing that they all contain a triangle of one colour.

So, we need a proof that can extend our knowledge beyond what is visualizable, by working its way, step by step, through the complexity.

Proof: Take one of the points, and call it O. Then of the five lines from that point to the others, at least three must have the same colour, say colour A. Consider the three points at the end of those lines. If any two of them are joined by a line of colour A, then they and O form an A-colour triangle. But if not, then the three points must all be joined by B-colour lines, so there is a B-colour triangle. So, there is always a single-coloured triangle.

There is nothing in this proof except what Aristotelian mathematical philosophy says there should be – no arbitrary axioms, no forms imposed by the mind, no constructions in Platonist set theory, no infinite sets, no dubious intuitions – only the necessary relations of simple structural universals and our certain insight into them, induced by following a chain of individually clear inferences. A written proof should communicate a sequence of insights to the reader, insights into necessary connections between the (quantitative or structural) universals dealt with by mathematics.[1] That is so irrespective of whether individual proofs can be fitted into an overarching deductive scheme of axioms and theorems, or whether they can be expanded into purely syntactic form.

The ancients and deductive proof

In the mid-fifth century BC, the Parthenon was rising over Athens, built according to the best geometrical principles. The construction lines are not visible on it, though they are on some other Greek temples.[2] Among the tragedians, artists, sophists and merchants creating Western civilization in the city below, a number of visitors from out of town brought an interest in what we would now call scientific questions. According to Parmenides of Elea, for example, thinking about the geometry of eclipses is enough for a convincing argument that the earth is round, even for someone fixed in one place on the earth's surface.[3]

A lesser-known genius who visited Athens was a certain Hippocrates of Chios.[4] He came to appreciate that geometry could be organized so that the complicated and less obvious propositions followed with strict logic from the simple and obvious ones. His project was perfected a century or more later in the *Elements* of Euclid, which was such a success that its predecessors have not survived. But one fragment of Hippocrates's work is left, enough to demonstrate his extraordinary brilliance. Being still at the 'bare hands' phase of mathematical development, it can be appreciated by anyone.

The fragment concerns the 'quadrature of lunes', that is, finding the area of a crescent-shaped figure as shown in the shaded portion at the top left of Figure 4.3.

The lune is bounded by two arcs of circles: the upper one AEB which is half of the circle with diameter ADB, and the lower one AFB which is a quarter of the circle with diameter AOC. Hippocrates proves, amazingly, that the area of the lune is exactly the same as the area of the right-angled triangle AOB (also shaded).

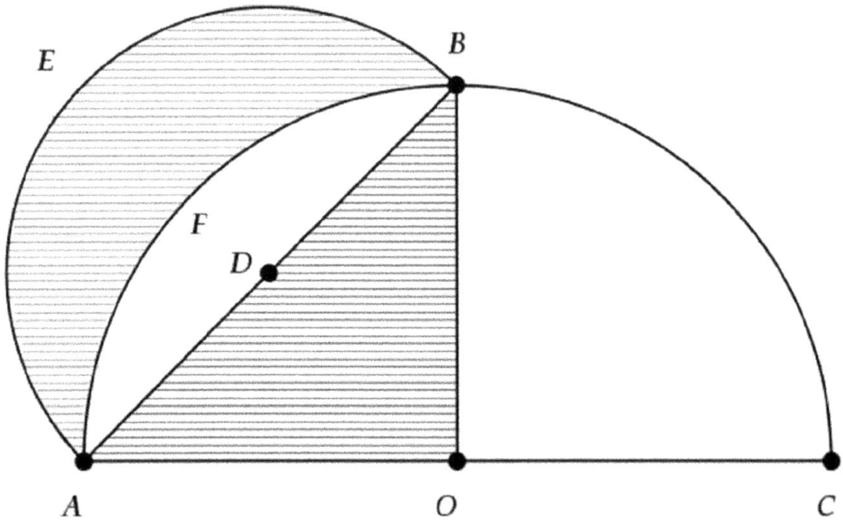

Figure 4.3 The Lune of Hippocrates.[5]

Note that there is no π in the answer, as there is in the formula for the area of a circle: Hippocrates does not use anything about the area of a circle, but proves directly that the two areas, one curved and one straight-sided, are equal. If the radius AO is one unit of length, both shaded areas are exactly half a square unit.[6]

The Greeks immediately realized there was something special about deductive proof as a way of acquiring knowledge. It is not like measuring a number of lunes and triangles and finding that, in all cases, the lune equals the triangle in area. The proof somehow gets to a deeper level of reality. It reveals not only what *is* so, but what *must* be so. And it allows us not only to know what must be so, but to understand *why* it must be so.

The results are not thereby cut off from physical reality, which is so if they are about a Platonic realm of ideal forms. If real lunes and triangles are drawn on paper, they are nearly equal in area; the more perfectly they are drawn, the nearer the drawn shapes approximate the exact parts of real space to which the proof applies, so the nearer they are equal in area.

There are plenty more exact and provable results where that came from, as laid out in the thirteen books of Euclid's *Elements* and many thousands of mathematical books and papers since then. The method of deductive proof generates endless outcomes, limited only by human interest and effort.

The Aristotelian-Euclidean model of science

The era of the Scientific Revolution inherited from the ancient world a model of mathematics as a deductive structure of propositions proved from self-evident truths. It was backed by the two greatest possible authorities, Aristotle and Euclid and was, in any case, open to inspection if one followed the proofs in Euclid oneself. Virtually all thinkers accepted it without question.

The essentials of the model were laid down in Aristotle's *Posterior Analytics*. A true or fully developed science should demonstrate its truths by syllogistic deduction from self-evident first principles, explaining why the truths of the science must be as they are. The first principles should be simple enough to be evident, after inquiry, to the pure light of reason, or *nous* (Latin *intellectus*), a divinely granted faculty of the soul capable of grasping necessities.[7]

If to modern ears that seems a promise too good to be true, for early modern thinkers it was confirmed by the existence and obvious success of Euclidean geometry, a science of space that seemed to conform exactly to Aristotle's model.[8] Euclid's *Elements* begins with twenty-three definitions, such as:

> 1. A *point* is that which has no part . . .
>
> 15. A *circle* is a plane figure contained by one line such that all the straight lines falling upon it from one point among those lying within the figure equal one another . . .

These are meant to be not arbitrary stipulations but contentful truths about real geometrical objects. Though Euclid omits any philosophical commentary, modern writers rightly emphasize that to appreciate the definition of a circle, for example, one needs an act of insight to understand that something uniformly round does, in fact, consist of all the points equidistant from a centre.[9] There follow five postulates or axioms, such as

> 1. To draw a straight line from any point to any point . . .
>
> 5. That, if a straight line falling on two straight lines makes the interior angles on the same side less than two right angles, the two straight lines, if produced indefinitely, meet on that side on which are the angles less than the two right angles.

The way in which the Fifth Postulate is posed is such as to make it appear as self-evident as possible, despite the worries, expressed since ancient times, as to whether anything involving lines 'produced indefinitely' could be truly self-evident. Indeed, behind the scenes there were considerable doubts among the best professional mathematicians as to whether Euclid's logic was watertight[10] but, from the perspective of outsiders, the Euclidean structure was monolithic and impregnable.

There follow a large number of propositions or theorems derived logically from the postulates, beginning with:

I.1 To construct an equilateral triangle on a given finite straight line.

These truths were taken to be necessary truths about the real space we live in, even though Aristotle, to some extent, adopts the Platonist idea that the geometer's lines and circles are assumed to be perfect and hence are idealizations of the real shapes of rulers and wheels.

The Aristotelian-Euclidean model after non-Euclidean geometry

The old view that the real space we live in is necessarily Euclidean proved to be false. The non-Euclidean geometries discovered in the nineteenth century are consistent and are possible shapes of space. It has come to be understood that it is an empirical question what shape space has – whether it is flat (Euclidean), infinitely extended, or infinitely divisible. Thus, Euclid's axioms and theorems cannot be necessary truths about space.

The discovery of non-Euclidean geometry has thus always been the main exhibit when there are attacks on the certainty of mathematics. It was central to a book once popular in 'mathematics for liberal arts' courses, Morris Kline's *Mathematics: The loss of certainty*, which represents the mathematics of the period 1830–1930 as a discipline lurching from crisis to crisis.[11] Paul Ernest's *Social Constructivism as a Philosophy of Mathematics* (1998) is typical of many in arguing that 'those assumptions that the community of mathematicians regard as self-evident in one era often become the focus of intense scrutiny and doubt in another era (e.g., the axioms of geometry before and after Kant, and the axioms of arithmetic before and after Peano)'.[12]

A significant mistake was made, but the nature of the mistake needs careful analysis. It is true that Euclidean geometry was once thought to be necessarily true of space and the existence of mathematically possible alternatives showed that the shape of space was an empirical question rather than a purely mathematical one. But that means that the mistake was about the physics of space, the philosophy of space and mathematical modelling: the mathematical model of Euclidean space may not fit actual space exactly, because of the nature of space. That is a bad mistake, but not exactly a mathematical mistake. It is more like mistakenly supposing that wheels and coins are perfectly circular and that π can be found to infinite accuracy by measuring them; that is a mistake about how accurately a certain mathematical structure is realized in a part of physical reality, rather than a mistake about the mathematics of circles. Euclid's theorems are all still true of the abstract structure, Euclidean geometry. That structure is realized in our space at our scale to a very close approximation, but possibly not exactly.

We understand now that geometry studies several mathematical structures, not just one. Mathematics studies all of them and physics measures which one applies; that is no different from arithmetic studying all numbers and the census department determining which one applies to the population.

A mistake was also made about epistemology, concerning self-evidence. Once the nature of space as physical is understood, it is clear that there can be no genuine self-evidence as to its shape. It cannot be necessary that space should extend infinitely or be infinitely divisible, so it is impossible that self-evidence should extend to those propositions. There were indeed long-running doubts on the self-evidence of Euclid's Fifth Postulate, from Euclid's own expression of it in a way that deliberately downplays its reference to indefinitely distant space, to Zeno of Sidon's observations that certain things did not follow from Euclid's axioms without further assumptions[13] and Saccheri's eighteenth-century project to 'free Euclid from all spot'.[14]

On the other hand, the possibility of discrete space was rarely considered, except by 'men of the world' like the seventeenth-century Chevalier de Méré (whose belief in atomic space was taken by Pascal as evidence of his incompetence in mathematics[15]) and a radical philosopher like Hume. That is certainly a major error, especially as Euclidean geometry was held up as the model of self-evidence. But again, the nature of the error is philosophical, not mathematical. Nor does it cast doubt on the other major epistemological fact, that it is easily perceivable and measurable that Euclidean geometry is a very accurate model of the structure of our space, in our region, at our scale. The mathematical results of Euclidean geometry are thus all directly true of our space (at that scale).

So, the claims of the 'loss of certainty' in mathematics are very much exaggerated. All the number theory in Euclid is still as true as it ever was. So is the geometry, if interpreted as being about a structure that is only approximately realized in our space.

It is also notable that there is no parallel to non-Euclidean geometries in number theory. There seem to be no alternative arithmetics in the same sense as there are alternative geometries. There are indeed many algebraic systems with some similarities to the numbers, such as the quaternions, but they are not regarded as *alternatives*, for example, as possible other ways to count, in the way that hyperbolic space is a genuine alternative to the Euclidean way of being spatial.

Proof in computer science: program verification

Since the time of Aristotle and Euclid, scientists and philosophers alike have been disappointed that the Aristotelian/Euclidean model of deduction from necessarily true premises has proved so rarely applicable outside pure mathematics. In natural sciences, it has been necessary to make do with an ersatz 'nomic' kind of necessity, which somewhere requires empirical evidence to establish laws of nature. We conclude with a few examples of more modern cases where genuine proof has escaped the narrow field of pure mathematics.

The late 1960s were the years of the 'software crisis', when it was realized that creating large programs free of bugs was much harder than had been thought. It was agreed that, in most cases, the fault lay in mistakes in the logical structure of the programs: there were unnoticed interactions between different parts, or possible cases not covered. One suggested remedy was that, since a computer program is a sequence of logical steps like a mathematical argument, it could be *proved* to be correct. The 'program verification' project has had a certain amount of success in making software error-free; mainly, it appears, by encouraging the writing of programs whose logical structure is clear enough to allow proofs of their correctness to be written. A great deal of time and money has been invested in this activity. But the question is, does the proof guarantee the correctness of the actual physical program that is fed into the computer, or only of an abstraction of the program? C. A. R. Hoare, a leader in the field, made strong claims:

> Computer programming is an exact science, in that all the properties of a program and all the consequences of executing it can, in principle, be found out from the text of the program itself by means of purely deductive reasoning.[16]

Some other authors explain why a program is not like a physical machine:

> By contrast [to hardware], a computer program is built from ideal mathematical objects whose behaviour is defined, not modelled approximately, by abstract

> rules. When an if-statement follows a while-statement, there is no need to study whether the if-statement will draw power from the while-statement and thereby distort its output, or whether it could overstress the while-statement and make it fail.[17]

The philosopher James Fetzer, however, argued that the program-verification project was impossible in principle.[18] Published not in the obscurity of a philosophical journal, but in the prestigious *Communications of the Association for Computing Machinery*, his attack threatened the livelihood of thousands. Fetzer's argument relied wholly on the gap between abstraction and reality.

It has been subsequently pointed out in addition that to predict what an actual program does on an actual computer, one needs to model not only the program and the hardware but also the environment, including, for example, the skills of the operator.[19] There can be changes in the hardware and environment between the time of the proof and the time of operation.[20] In addition, the program runs on top of a complex operating system, which is known to contain bugs. Plainly, certainty is not attainable about any of those matters.

But there is some mismatch between those undoubtedly true considerations and what was being claimed. Aside from a little inadvised hype, the advocates of proofs of correctness had admitted that such proofs could not detect, for example, typos.[21] And, on examination, the entities Hoare had claimed to have certainty about were, while real, not unsurveyable systems including machines and users, but written programs.[22] That is, they are the same kind of things as published mathematical proofs.

For comparison: If a mathematician says, in support of his assertion, 'my proof is published on page X of volume Y of *Inventiones Mathematicae*', one does not normally say – even a philosopher does not normally say[23] – 'your assertion is attended with uncertainty because there may be typos in the proof', or 'perhaps the Deceitful Demon is causing me to misremember earlier steps as I read later ones'. The reason is that what the mathematician is offering is not, in the first instance, absolute certainty, in principle, but necessity. That is how his assertion differs from one made by a physicist. A proof offers a necessary connection between premises and conclusion. One may extract practical certainty from that, given the practical certainty of normal sense perception, but that is a separate step. That is, the certainty offered by mathematics does depend on a normal anti-scepticism about the senses, but it removes, through proof, the further source of uncertainty found in the physical and social sciences, arising from the uncertainty of inductive reasoning and of theorizing.[24] Assertions in physics about a particular case have two types of uncertainty: that arising from the measurement and observation needed to check that the theory applies to the case, and that of the theory itself, arising from the non-necessary, merely nomic, nature of physical reality. Mathematical proof has only the first.

It is the same with programs. While there is a considerable certainty gap between reasoning and the effect of an actually executed computer program, there is no such gap in the case Hoare was considering, the unexecuted program. A formalized proof (in, say, the predicate calculus) is a sequence of steps exhibiting the logical connection between formulas, and checkable by humans (if it is short enough). Likewise, a computer program is a logical sequence of instructions, the logical connections among which are checkable by humans (if there are not too many).

Thus, Fetzer's arguments fail to show that computer programs cannot be proved correct. They can have provability in the same maximally strong sense as mathematical proofs.

One feature of programs that is inessential to this reply is their being textual. So, one line taken by Fetzer's opponents was to say that not only could programs be proved correct but so could machines. Again, it was admitted that there was a theoretical possibility of a perceptual mistake, but this was regarded as trivial, and it was suggested that the safety of, say, a physically installed railway signalling system could be assured by proofs that it would never allow two trains on the same track, no matter what failures occurred.[25] The advertisement that said:

> VIPER is the first commercially available microprocessor with both a formal specification and a proof that the chip conforms to it

was felt by the experts to be a danger to the gullible public, but not an impossibility in principle.[26] An aggrieved purchaser began legal action on the grounds that the proof was not complete, but the bankruptcy of the plaintiff unfortunately prevented this interesting philosophical debate from being pursued in the courts.[27]

Proof in the contingent, explanation, and mathematical modelling

Proofs about computer hardware, about an arrangement of real bridges, about a Euclidean geometry that space has but might not have – how is it possible for proof to apply to the physical world when physical properties are contingent?

That is not like a pure mathematical proof in number theory, which proves truths about a wholly necessary subject matter.

Aristotelian realist philosophy of mathematics has a simple answer to that question. Mathematics is the science of structure. Machines, systems of bridges, and space have (contingently) certain structures. Mathematics proves things about those structures – about the necessary interconnections between parts and the consequences of having such structures. Therefore, the mathematical proofs show something necessary about the properties of the physical things themselves. As d'Alembert puts it with respect to mechanics:

> No theory could have allowed us to find the law that heavy bodies follow in their vertical fall, but once this law is found through experience, all that belongs to the movement of heavy bodies, whether rectilinear or curvilinear, whether inclined or vertical, is found entirely by theory.[28]

Thus, there can be mathematical explanations – complete and necessary explanations – about physical things, even though the existence and properties of the things themselves are contingent. The physical things do not exist necessarily, nor do they have their physical properties necessarily, but some of the interconnections between those properties are necessary and subject to mathematical explanation in full.

A simple but applicable example is the geometrical explanation of the Coriolis Effect, which is responsible for hurricanes in the northern hemisphere swirling anti-clockwise and those in the southern hemisphere clockwise. Two contingent but very general facts are needed to get the explanation under way:

1. Newton's first law, that a moving object tends to move in a straight line with constant velocity.
2. The fact that gravity pulls particles such as wind particles towards the centre of the earth, so that they stay near the surface of the globe.

Now consider a free particle such as a wind particle, travelling north in mid-latitudes of the northern hemisphere (the small circle in Figure 4.4).

The earth rotates towards the east (which is why the sun rises in the east). By Newton's first law, the particle tends to travel tangentially north, along the solid arrow (that is, actually travelling east with the earth but, in addition, travelling north). But tangentially is off the planet, so gravity pulls it back down towards the surface – that is, onto a parallel of latitude with smaller radius than further south where it started. Thus, the easterly component of motion it retains means that it 'gets ahead' of the earth's surface, that is, the particle is deflected (from the

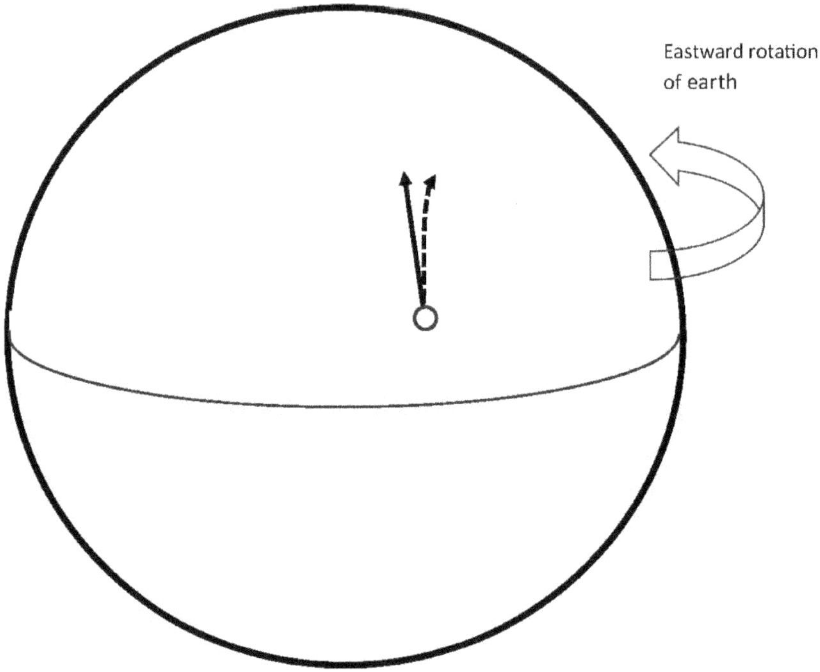

Figure 4.4 The Coriolis effect on a particle travelling north in the northern hemisphere.

point of view of the earth's surface) to the east (that is, right as seen from its direction of flow (dashed curve).

Similar reasoning will show that a particle moving south in the northern hemisphere is deflected right (west). Somewhat similar but more difficult reasoning shows that particles travelling east or west are also deflected right.

Hurricanes are caused by rising air in their centre, so flow is inwards toward the centre. Inward-flowing air all deflected to its right creates a counter-clockwise global effect. (And the opposite in the southern hemisphere.)[29]

The example provides a paradigm of how mathematics can explain physical phenomena. The physical phenomena are not necessary, absolutely speaking, since they involve some contingent principles such as Newton's first law or the Euclidean shape of space. But in complex examples, the weight on the explanation rests on the connections between those contingent principles and the phenomena. Mathematics can prove and explain the necessity of those connections without remainder.

We conclude with the promised proof of Euler's result that it is impossible to walk over all seven of the bridges of Königsberg once and once only.

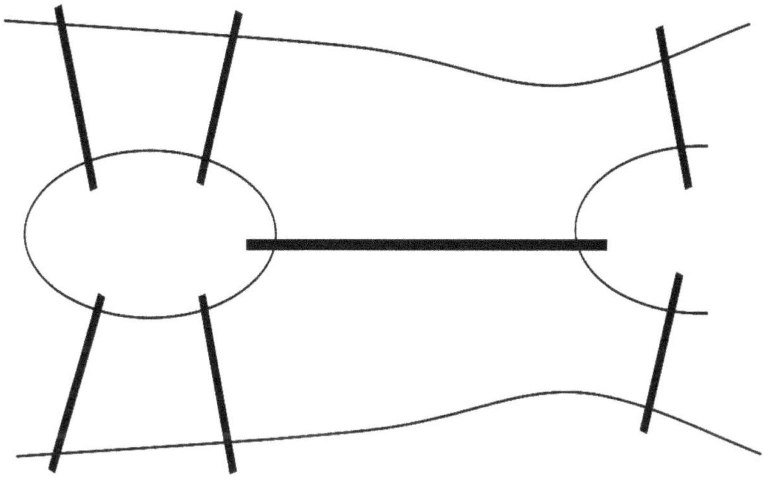

Figure 4.5 The bridges of Königsberg.

Euler's own proof is readable but a little clumsy. The usual modern proof is simpler. If a path moves into and out of a land area, it uses up two of that area's bridges – one to enter and one to leave. So, if there exists a path going over all the bridges once and once only, the number of bridges leaving any land area must be even (except possibly the two land areas where the path begins and ends). But in the Königsberg case, the number of bridges leaving every land area is odd (5 in one case, 3 in the others). Therefore, there cannot exist a path going over all the bridges once and once only. The proof has explained perfectly why things must be so in physical reality.

Notes

1 A textbook on proof from this point of view is James Franklin and Albert Daoud, *Proof in Mathematics: An Introduction* (Sydney: Kew Books, 2011).

2 Lothar Haselberger, 'The Construction Plans for the Temple of Apollo at Didyma,' *Scientific American* 253, no. 6 (1985): 126–32.

3 Otto Neugebauer, *A History of Ancient Mathematical Astronomy* (Berlin: Springer, 1975), 576.

4 Confusingly, not the same person as Hippocrates of Cos, the medical writer after whom the Hippocratic Oath is named.

5 Taken from https://upload.wikimedia.org/wikipedia/commons/e/e0/Lune.svg.

6 The original text of his proof is given in this footnote in case some readers wish to skip it. 'He started with, and laid down as the first of the theorems useful for the purpose, the proposition that similar segments of circles have the same ratio to one another as the squares on their bases. And this he proved by first showing that the squares on the diameters have the same ratio as the circles. After proving this, he proceeded to show in what way it was possible to square a lune the outer circumference of which is that of a semicircle. This he effected by circumscribing a semicircle about an isosceles right-angled triangle and a segment of a circle similar to those cut off by the sides. Then, since the segment about the base is equal to the sum of those about the sides, it follows that, when the part of the triangle above the segment about the base is added to both alike, the lune will be equal to the triangle.' Text of Simplicius, translated in Thomas L. Heath, *A Manual of Greek Mathematics* (Mineola NY: Dover, 2003), 121–32; a more accessible explanation can be found in W. S. Anglin, *Mathematics: A Concise History and Philosophy* (New York: Springer, 1994), 52.

7 David Bronstein, *Aristotle on Knowledge and Learning: The Posterior Analytics* (Oxford: Oxford University Press, 2016).

8 Richard D. McKirahan, *Principles and Proofs: Aristotle's Theory of Demonstrative Science* (Princeton: Princeton University Press, 1992), ch. 12.

9 Bernard Lonergan, *Insight: A Study of Human Understanding,* (5th edn, Toronto: Toronto University Press, 1997), ch. 1.

10 Vincenzo de Risi, 'The Development of Euclidean Axiomatics: The Systems of Principles and the Foundations of Mathematics in Editions of the *Elements* in the Early Modern Age', *Archive for History of Exact Sciences* 70 (2016): 591–676.

11 Morris Kline, *Mathematics: The Loss of Certainty* (New York: Oxford University Press, 1980), 78–88.

12 Paul Ernest, *Social Constructivism as a Philosophy of Mathematics* (Albany NY: SUNY Press, 1998), 26.

13 Proclus, *Commentary on the First Book of Euclid's Elements*, trans. G. Morrow, 2nd edn (Princeton: Princeton University Press, 1992), 156, 169–71.

14 G. Saccheri, *Euclides ab Omni Naevo Vindicatus* (Milan, 1733); M.J. Greenberg, *Euclidean and Non-Euclidean Geometries: Development and History* (4th edn, New York: Freeman, 2007), ch. 5.

15 Pascal to Fermat, July 29, 1654, in D. E. Smith, *A Source Book in Mathematics* (New York: Dover, 1959), II, 552.

16 C. A. R. Hoare, 'An Axiomatic Basis for Computer Programming', *Communications of the Association for Computing Machinery* 12 (1969): 576–80.

17 Richard Stallman and Simson Garfinkle, 'Against Software Patents', *Communications of the Association for Computing Machinery* 35 (1992): 17–22 and 121, at 19.

18 James H. Fetzer, 'Program Verification: The Very Idea', *Communications of the Association for Computing Machinery* 31 (1988): 1048–63.

19 Jon Barwise, 'Mathematical Proofs of Computer System Correctness', *Notices of the American Mathematical Society* 36 (1989): 844–51.

20 Meir M. Lehman, 'Uncertainty in Computer Application', *Communications of the Association for Computing Machinery* 33 (1990): 584–86.

21 David Gries, *The Science of Programming* (New York: Springer, 1981), 5

22 C. A. R. Hoare, 'Programs are Predicates', in *Mathematical Logic and Programming Languages*, (eds) C. A. R. Hoare and J. C. Shepherdson (Englewood Cliffs, NJ: Prentice Hall, 1985), 141–55.

23 Unless he is Hume: *Treatise of Human Nature,* I.IV.i.

24 A similar argument is made of computer-assisted proofs of theorems in Tyler Burge, 'Computer Proof, A Priori Knowledge, and Other Minds', *Philosophical Perspectives* 12 (1998): 1–37, section II.

25 H. M. Müller, letter, in *Communications of the Association for Computing Machinery* 32 (1989): 506–508; cf. Hoare, 'Programs are Predicates', note 56.

26 John Dobson and Brian Randell, 'Program Verification: Public Image and Private Reality', *Communications of the Association for Computing Machinery* 32 (1989), 420–422; Donald MacKenzie, 'The Fangs of the VIPER', *Nature* 352 (1991): 467–8; later developments in Christoph Kern and Mark R. Greenstreet, 'Formal Verification in Hardware Design: A Survey', *ACM Transactions on Design Automation of Electronic Systems* 4 (1999): 123–93.

27 Donald MacKenzie, 'Computers, Formal Proofs and the Law Courts', *Notices of the American Mathematical Society* 39 (1992): 1066–69.

28 Jean d'Alembert, article 'Expérimentale', in *Encyclopédie, ou Dictionnaire Raisonné des Sciences, des Arts, et des Métiers*, ed. D. Diderot and J. d'Alembert, 17 vols (Paris, 1751–65), 4: 300.

29 Henry M. Stommel and Dennis W. Moore, *An Introduction to the Coriolis Force* (New York: Columbia University Press, 1989); an intuitive introduction in David J. Van Domelen, 'Getting around the Coriolis force' (1996). Available at: https://www.eyrie.org/~dvandom/Edu/newcor.html

Part Three

Logical Necessities, Probabilistic and Deductive

5

Logical probability and the relation of evidence to hypothesis

As in mathematics, so in the relation of evidence to hypothesis, the human mind can have insights into absolute necessities. Just as we can understand why $2 \times 3 = 3 \times 2$, we can understand that some bodies of evidence give strong support to a hypothesis, others tell against it, and yet others are irrelevant. Juries can be right or wrong in evaluating evidence of guilt as beyond reasonable doubt, at least in cases where the evidence clearly tends one way or the other. We can understand why the order of presentation of evidence for a hypothesis is irrelevant to how strongly the evidence supports the hypothesis.

Two kinds of probability: stochastic versus logical/epistemic

Probability is an inherently confusing subject. A principal problem is that the word refers to two quite different concepts, and failure to keep them separate has caused endless trouble.[1] First, there is *factual* or *stochastic* or *aleatory* probability, dealing with chance set-ups such as dice-throwing and coin-tossing, which produce characteristic random or patternless sequences of outcomes. Almost always, in a long sequence of coin tosses, there are about half heads and half tails, but the order of heads and tails does not follow any pattern. Here is an example of a series of heads and tails from throwing a real coin:

THHHTHHHHHTTHHHTHTHHHHTTTTHHHTHHHTTHTHTHTTHHHTHT

Plainly, the lack of pattern in the outcomes is an objective matter of the way the world is and has nothing to do with any mentalistic or psychological or logical concepts such as uncertainty or predictability. (Though patternlessness may *imply* unpredictability, it is patternlessness that comes first.) There are objective tests of randomness for long sequences. For example, in a random sequence of Hs and Ts, there should be about half Hs and half Ts, about a quarter of the pairs should be each of HH, HT, TH and TT, and so on. Deterministically generated sequences like the digits of π or the 'random numbers' generated by a computer can possess the same kind of patternlessness, so it is not a property necessarily associated with anything stochastic.

Notice too that the phenomenon of patternlessness in a series of coin tosses does not disappear as we learn more about the microphysics of the situation. It may be that dice throws are fully deterministic and that if we could measure the initial positions and strength of the throw of the coins, we could predict the outcome of each throw. Even if we did that with each throw, we would not have explained why about half the throws were heads and half tails, nor why they came up in a patternless sequence. Explaining this needs to refer to physical probabilities, whatever account may be given of those.

It would not be true to say, though, that probability *just is* the relative frequency of outcomes. The probability of heads is a property of the coin and its throwing mechanism, the *cause* of the relative frequency of outcomes. There is always the possibility of mismatch between the probability of heads and the actual frequency of heads in the throws. Indeed, there is almost always some mismatch: a perfectly fair coin thrown 1,000 times is very unlikely to produce exactly 500 heads. Even gross mismatches between probability and outcomes are possible: one can throw a fair coin indefinitely and always throw heads (though it is not likely). That is what makes factual probabilities *probabilities* – they can always fail to be realized. Thus, a factual probability (tendency, bias, propensity) must be regarded as a hidden cause in a repeatable mechanism, which normally will be but sometimes may not be realized when the mechanism is repeatedly run.

In contrast to factual or stochastic probability, there is another kind, *logical* or *epistemic* probability, concerned with how well-supported a conclusion is by a body of evidence. A concept of logical probability is employed when one says that, on present evidence, the steady-state theory of the universe is less probable than the Big Bang theory, or that an accused's guilt is proved beyond reasonable doubt though not absolutely certain. How probable a hypothesis is, on given evidence, determines the degree of belief it is rational to have in that hypothesis, if that is all the evidence one has that is relevant to it. The concept is crucial to theory evaluation of science.

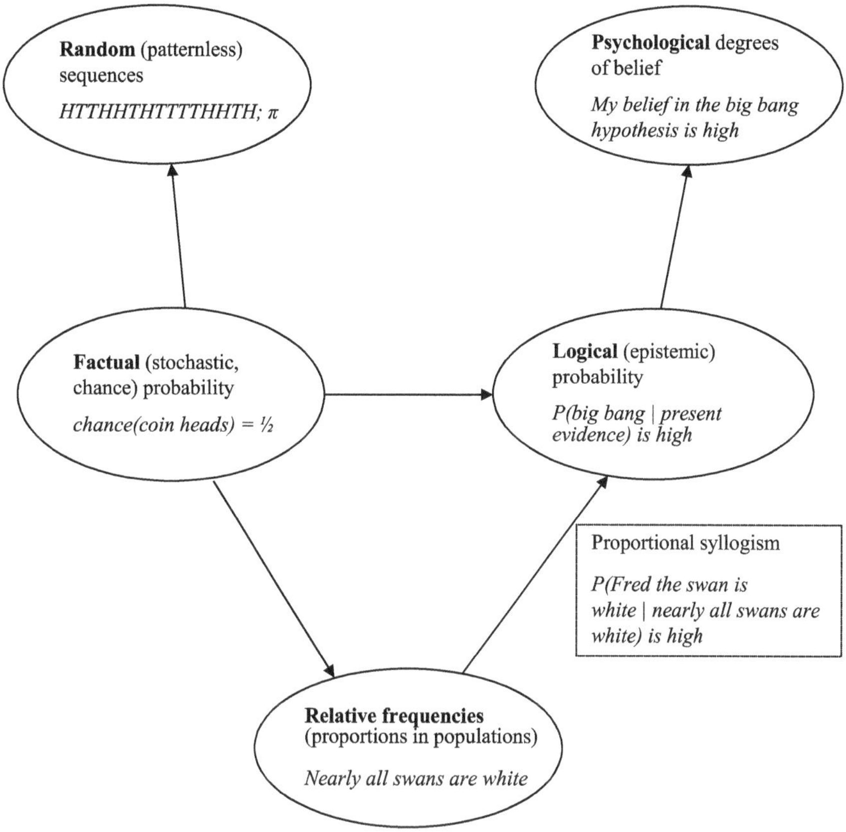

Figure 5.1 Kinds of probability and their interrelations.

By way of clarification, here is a map of the territory covered by the word 'probability'. $P(h|e)$ is used to mean 'The (logical) probability of hypothesis h given evidence e'.

Logical probabilities cannot be regarded as the same thing as psychological degrees of belief: degrees of belief *ought* to conform to probabilities – what is *really likely* on the evidence – but people being what they are, they often do not. The famous psychological results of Kahneman and Tversky showed that humans' deductive as much as non-deductive reasoning sometimes fails to conform to sound principles – for example, by being sensitive to the order of presentation of evidence.[2] Failing to distinguish between the logic of science and actual degrees of belief would lead to the philosophy of science being replaced by its history and sociology, a failing whose widespread effects in post-Kuhnian philosophy of science are only too evident.[3]

Logical probability cannot be reduced to factual. There is no chance, propensity or frequency involved in, say, whether design arguments for the existence of God are convincing. There is no ensemble of universes throwing themselves, with some relative frequency of divinity in them, but only the question of the conclusiveness of the non-deductive inference from apparent design to an intelligent designer. Conversely, factual probabilities cannot be about knowledge, belief or logic: the characteristic patternlessness observed in the sequences of dice throws is a physical fact independent of human knowledge and requires a physical cause.

While factual probabilities are inherently numerical, being either proportions or tendencies to produce a definite proportion in outcomes, it is not obvious whether logical probabilities should be assigned numbers, sometimes or always. We will consider that in a later section on the formalization of logical probability.

Of course, there are connections between the two kinds of probability, distinct though they are. There are also connections between probabilities and relative frequencies. The most important connection is the proportional syllogism, which states the logical probability of an outcome when the evidence is a relative frequency (discussed below). A second connecting principle is what the philosopher David Lewis called the Principal Principle, which does the same when the evidence is a statement of factual probability. Some principle is needed to connect any such physical quantity as a factual probability with one's rational degree of belief. If, as a matter of physics, the chance of a die coming up 6 is one-sixth, what should one rationally believe about the next throw of the die? The physics may cause the die to come up 6 about one-sixth of the time (usually) but what one's degree of belief ought to be is a matter of logic, not physics. The Principal Principle says that, on this evidence, one's degree of belief in the proposition that the next throw will land 6 ought to be one-sixth.[4] That is a statement of logical probability, and one indispensable for making an inference in stochastic situations.

Evidence evaluation in law, science, and philosophy

Law has always been forced to deal with conflicting evidence – of witnesses, documents, generally known facts and rumours. That is why the law of evidence was the central topic in the history of the development of probability, in the centuries before Pascal and Fermat began applying mathematical techniques to it.[5] More will be said in the last chapter below.

The theory of logical probability gives the law of evidence a foundational story about the nature of evidence: a jury or judge must evaluate the logical relation between the evidence presented in court and the conclusion in dispute in the case.

The theory of logical probability, for example, explains the justification for exclusionary rules of evidence. When the Evidence Act 1995 says:

> The court may refuse to admit evidence if its probative value is substantially outweighed by the danger that the evidence might
>
> (a) be unfairly prejudicial to a party; or
> (b) be misleading or confusing.[6]

or

> In a criminal proceeding, the court must refuse to admit evidence adduced by the prosecutor if its probative value is outweighed by the danger of unfair prejudice to the defendant.[7]

such concerns only provide a reason for reform if one has an objective conception of evidence, according to which there can be a mismatch between court practices and what the court should, as a matter of objective fact, believe in the evidence. The theory that probability is logic has a clear story about what it means. 'Probative value' is the logical degree to which the evidence really does support the conclusion, while 'prejudice', 'misleading' and 'confusing' are ways in which people's degrees of belief might be caused to deviate from what they ought to be, on the evidence. The possibility of – indeed the persistent and predictable tendency towards – deviation between logic and psychology is the reason for having the exclusionary rule. It is hard to see an alternative theory of evidence substantially different from logical probability making sense of that.

The same applies to experimental and observational evidence in science. The true lover of science will revel first in the many low-level empirical generalizations that summarize and give shape to our long perceptual experience: 'All ravens are black', 'Tonight's stars form much the same patterns as yesterday night's', 'Banana peels are normally slippery', 'Virgins don't have children', 'It's in spring that the wheat comes up', 'Lithium is effective against bipolar disorder', 'Tossed coins come up heads about half the time'. Those are established by observation, followed by inductive inference to establish a generalization from the finite number of instances surveyed. (The rationality of inductive inference will be considered below.) More theoretical results can be logically confirmed by their predictions, as also described below.

And it is not only scientifically sophisticated adults who reason probabilistically. There has also been a growing realization among developmental psychologists that babies are powerful 'Bayesian' reasoners, that is, their reasoning conforms to the laws of probability. Six-month-old infants show surprise if a box, which they know contains mostly pink and a few yellow balls, reveals a sample that is mostly yellow, whereas they are not surprised when the sample is mostly pink. That is, they know implicitly that the composition of a sample is likely to approximately match the composition of the population.[8] Twelve-month-olds have much more sophisticated abilities to integrate such frequency perceptions with other knowledge, such as expectations about the motion of objects.[9] Regarding infants as 'Bayesian ideal observers' has proved to be predictive of baby behaviour in such studies. Since these abilities are acquired even before infants have learned to speak, it is clear that humans have pre-linguistic abilities to respond to and reason about probabilities.

The deductivist error

Given the crucial role played by logical probability or non-deductive logic in ordinary life, science and law, it is surprising that it is not discussed more explicitly, and that an implicit 'deductivism' (the thesis that all logic is deductive) is in many circles almost orthodoxy. Deductivism is a thesis very corrosive of rationality.[10] If one believes that, it will be impossible to resist the conclusion that if theories are not maintained for deductively good reasons, they must be maintained for political reasons like patronage or state propaganda (or maybe psychological ones such as ego defences). As David Stove showed, deductivism was behind both Hume's scepticism about induction and the central theses of Karl Popper, in his *Logic of Scientific Discovery,* and Thomas Kuhn in *The Structure of Scientific Revolutions,* that science could never genuinely advance through the accumulation of evidence making theories more probable.[11] Those two books dominated philosophy of science in the mid- and late-twentieth centuries. They opened the way to even more irrationalist currents, which replaced philosophy of science with the sociology of science – and why not, if scientific beliefs have only causes and not reasons? Fortunately, recent trends have shifted towards an objective Bayesian or logical probabilist philosophy of science.[12] But in the theory of both science and law, there is still a long way to go before logical principles of probability are central to discussions of reasoning in those fields.

Two basic principles: confirmation by consequences and the proportional syllogism

It is all very well to say that logical probability exists and can be intuited in certain cases, but we will not understand how to actually determine if theories are well supported unless we understand what the principles of logical probability are.

There are two basic principles. They are easy to understand and agree with how everyone reasons, but they have a strangely low profile in discussions of rationality and even of probability. They are the confirmation of theories by their consequences, and the proportional syllogism.

The principle of confirmation by consequences says that if a theory has (logical) consequences and those consequences are found to be true, then the theory is confirmed, meaning that it is more probable than it was to begin with. While few have talked about this principle explicitly, George Polya rightly called it the 'fundamental inductive pattern'.[13] He expresses it as:

q is a (non-trivial) consequence of hypothesis *p*
q is found to be true
So, *p* is more likely to be true than before.

In short, 'Theories are confirmed by their consequences or predictions.' As described in the introduction above, a tribe that did not form their beliefs according to that principle would soon be extinct.

In other familiar examples: The detective reasons that if the butler did it, the knife should be behind the sofa. The knife is found behind the sofa, so the theory that the butler did it is more probable. Einstein's theory of general relativity predicted hitherto unexpected bending of light from distant stars; an expedition went out to West Africa and observed it, so relativity was confirmed.[14]

As in these cases, it is initially unlikely predictions that best confirm theories. 'Consequences' often means predictions, but not always. For example, if a conjecture in pure mathematics implies some known result, that confirms the conjecture. (Conjectures in mathematics are discussed below.)

The second basic principle of logical probability is the 'proportional syllogism', also called the statistical syllogism or direct inference. 'The syllogism' refers to the form of deductive inference, discussed since the ancient Greeks. In a classic example:

All men are mortal
Socrates is a man
Therefore, Socrates is mortal.

If the two premises are known to be true, it follows with certainty that the conclusion must also be true.

So, consider this – apparently quite similar – argument, called the 'proportional syllogism':

99% of men are mortal
Socrates is a man
Therefore, Socrates is mortal.

If the premises are known to be true, it is certainly *possible* that the conclusion is false. So, it is not a deductive form of inference. Nevertheless, the premises provide good reason to believe the conclusion. Nearly all men are mortal, and what could be better evidence that a particular one is? That sort of reasoning is everywhere in life: nearly all flights arrive safely, so one can relax on take-off; most people with one's disease improve with a certain drug, so it is a good idea to take it.

Note that the premises of the argument do not include: "Socrates is a randomly chosen man" or anything like that. Of course, if there is further evidence about Socrates, such as that he is a special friend of the gods, then it might not be reasonable to conclude that he is probably mortal. But that changes the premises, so is a different argument. It does not suggest there is anything wrong with the original argument, with the given premises.

The proportional syllogism raises a fascinating and difficult problem, the reference class problem, which arises from the fact that an individual is a member of several classes and so several proportional syllogisms apply to it. How can the evidence from them be combined?

It is all very well to argue:

Tex is a Texan and 90% of Texans are rich, so Tex is very probably rich.

But what if Tex is a member of several classes, with differing frequencies of wealth? Suppose the evidence is:

Tex is a Texan philosopher and 90% of Texans are rich and 10% of philosophers are rich.

What should one then think about the probability of Tex being rich?[15]

Plainly, the problem will appear in any case where there may be doubt as to what class containing an instance is most relevant to determining its probability of having some attribute. Philosophers, as is their way, have written at length on the ubiquity and difficulty of the problem, without offering a solution.[16] Yet humans have excellent intuition on combining disparate pieces of evidence, assigning (approximately) the correct weight to each. It is possible to understand how to decide the correct weight and imitate the intuitive human procedure, though the answer will not be included here.[17]

Formalizing logical probability: objective Bayesianism with imprecise probabilities

Those two intuitively strong principles – the confirmation of theories by consequences and the proportional syllogism – fit into a more general and formal theory known as (objective) Bayesianism with maximum entropy theory.

Bayesianism models the relation of evidence to hypothesis by means of the familiar axiom system of conditional probability: if h (for 'hypothesis') and e (for 'evidence') are two propositions, $P(h|e)$ ('the probability of h given e') is a number between 0 and 1 inclusive expressing the degree to which h is supported by e, which satisfies the two axioms:

$$P(\text{not-}h|e) = 1 - P(h|e)$$
$$P(h'|h \,\&\, e) \times P(h|e) = P(h|h' \,\&\, e) \times P(h'|e)$$

(The second axiom is a version of the Bayes Theorem that gives Bayesianism its name.)

However, that is something of an over-idealization of the true situation. Many Bayesians do believe that there is, in principle, a number (between 0 and 1) expressing the (logical) probability $P(h|e)$ of any given hypothesis h on any given body of evidence e; indeed, that might be called the most orthodox Bayesian position. But Keynes, whose *Treatise on Probability* of 1921[18] first clearly expressed the theory of logical probability, believed it was impossible to give every probability a number; certainly, it seems both impossible and pointless to debate precise numbers for, say:

P(Marilyn Monroe was murdered by the CIA | The moon is made of green cheese.)

There is no logical relation between 'evidence' and conclusion, so there is no point looking for a number to express it.[19] More generally, in situations where the evidence is as imprecise as in most legal cases, the positing of an in-principle precise number does no real theoretical work; for example, if it is clear on a large mass of evidence that guilt is almost certain but that a precise numerical probability, even if it existed, could not be computed, one is really working with an imprecise probability.

It is sufficient for objective Bayesianism that it is sometimes intuitively evident that some hypotheses, on some bodies of evidence, are highly likely, almost certain, or virtually impossible.

Sometimes it is indeed natural to assign an exact number, as in simple cases of the proportional syllogism. But in more complex cases where the evidence is itself imprecise or bears only remotely on the conclusion, it is harder to say. The legal system has strongly resisted assigning even an approximate number to the standard of 'proof beyond a reasonable doubt'.[20]

One might toy with the idea of accepting some minimal objective, but qualitative, notion of evidential relevance, while hoping to avoid taking on the full superstructure of Bayesian theorems. That is a vain hope – something like trying to accept mathematicians' advice on the addition of single-digit numbers while avoiding their excessively complex procedures for adding double-digit numbers. Once one has accepted the confirmation of theories by their consequences, one will be hard put to avoid accepting the next theorem, Polya's 'verification of an improbable consequence':

> q is a (non-trivial) consequence of hypothesis p
> q is very improbable in itself
> q is found to be true
> So, p is much more likely to be true than before.[21]

After agreeing to a number of such intuitively plausible theorems, one is on a slippery slope to full-blooded objective Bayesianism, according to which numbers, when applicable, do satisfy the axioms. There will be little motivation for wishing to avoid the more complex theorems, whose acceptance will make one a true believer.

It will be observed that the two Bayesian axioms state relations between probabilities, such as that between $P(\text{not-}h|e)$ and $P(h|e)$, but do not in themselves permit the actual numerical value of probabilities to be calculated. To do that, some further principle is needed, usually the proportional syllogism. The 'principle of maximum entropy', promoted by 'Maxent Bayesians' led by E. T.

Jaynes,[22] is a generalization both of the proportional syllogism and of the 'principle of insufficient reason' which assigns equal probability to symmetrical cases, such as assigning probability 1/6 to each face of a die in the absence of any evidence of bias.

The principle of insufficient reason has often been claimed to lead to paradoxes when different symmetries lead to different assignments of prior probabilities. Those are just cases of the problem of different reference classes, as discussed above in the case of the proportional syllogism. If different symmetries bear on a case, that is an objective feature of the evidence and the evidence from each of them needs to be combined with the correct weight of each – which may not be easy.[23]

Evaluating the strength of conjectures in pure mathematics

The logical character of probabilistic reasoning and the relation of hypothesis to evidence is best shown by its applicability in pure mathematics. The subject matter of mathematics is necessary, so there is no room for considerations about laws of nature, the uncertainty of observations, factual probabilities, or other such things that complicate probabilistic reasoning in natural science. In a mathematical case of a conjecture supported by numerical evidence, there is only the evidence, the conjecture and the logical relation between them.

But since mathematics is a science with a subject matter, the usual probabilistic methods of science work in it. So, in addition to methods special to mathematics such as proof, ordinary scientific methods such as experiment, conjecture and the confirmation of theories by observations work just as well in mathematics as in science.

Probabilistic reasoning is found in mathematics in those areas where mathematicians consider propositions that have not yet been proved. They are of two kinds. First, there are those that any working mathematician deals with in preliminary work before finding the proof for them, or indeed before finding the theorems to be proved. Second are the long-standing conjectures that have been written about by many mathematicians but which have resisted proof.[24] The most famous currently unsolved problem is the Riemann Hypothesis, which is believed by most competent mathematicians to be almost certainly true.[25]

A particularly straightforward and illuminating example of probabilistic reasoning, in fact of simple induction, in mathematics, is:

The first million digits of π are random
So, the second million digits of π are random.

The number π has the decimal expansion

3.1415926535897932384626433832795028841971693993751058209749445923
0781640628620899862803482534211706798214808651328230664709384460955058223172535940812848111745 02 . . .

There is no visible pattern in these digits. They appear 'random' in the sense of 'patternless'. (They are, of course, not random in the sense of 'stochastically generated' like the outcomes of dice throws.) That initial intuitive impression can be confirmed by statistical tests that show there are about the same number of each digit, in the long run, that there are about the same number of runs up (like 159) and down (like 653), and so on.[26] The sequence of digits exhibits complete patternlessness or randomness. The first million digits have long been calculated (calculations have reached beyond one hundred trillion digits).

Computer calculations applying statistical tests of randomness can confirm they are indeed random. It can then be argued inductively that the second million digits will likewise exhibit no pattern. This induction is a good one (indeed, everyone believes that the digits of π continue to be random indefinitely, though there is no proof), and there seems to be no reason to distinguish the reasoning involved here from that used in inductions about flames or ravens. But the digits of π are the same in all possible worlds, whatever natural laws may hold in them or fail to. Any reasoning about it is also rational or otherwise, regardless of any empirical facts about natural laws. Therefore, induction must be rational, independently of whether there are natural laws or whether nature is uniform.

In mathematics, there can be no confusion over natural laws, the regularity of nature, approximations, propensities, the theory-ladenness of observation, pragmatics, scientific revolutions, the social relations of science or any other red herrings. We see the absolutely necessary core of scientific inference laid completely bare, with only the hypothesis, the evidence and the logical relations between them.

The Williams-Stove justification of induction

An exceptionally interesting application of logical probability is the justification of induction using the proportional syllogism. *Prima facie*, inductive (or sample-to-population) reasoning like 'All observed ravens have been black, therefore the next raven will probably be black' is of a different and less logically secure nature than a proportional syllogism (population-to-sample) such as 'Nearly all ravens in England are black so the next English raven to be observed is probably black'. But an argument published in *The Ground of Induction* (1947) by the American philosopher D. C. Williams[27] and defended by David Stove in *The Rationality of Induction* (1986)[28] justifies induction in terms of the proportional syllogism.

The argument considers the case of a large box full of black and white balls in some unknown proportion and taking all the large samples from it and counting the proportion of blacks to whites in each sample. The argument (a proportional syllogism) is:

> The vast majority of large samples approximately match the population in composition.
> This is a large sample.
> Therefore, this matches the population.
> That is, the population matches the sample.

As with any proportional syllogism, the conclusion is (logically) probable but not certain, given the premises. If it is probable on the evidence that the population matches the sample, then inferring from sample to population – that is, induction – is justified, and justified for purely logical reasons. The first premise, 'The vast majority of large samples approximately match the population in composition', is a truth of pure mathematics. A logical justification of induction like this implies, for example, that induction works in all possible worlds, ordered or otherwise. It might at first seem that induction does not work in chaotic, lawless worlds, but in them the induction 'chaos, so more chaos' is sound. While there could be a world in which one's own inductions are regularly frustrated, for example by the machinations of the devil, a world in which inductions in general do not mostly work has not been found. After all the scepticism directed at induction in the last few centuries, it is surprising to find that its rationality does not depend on any contingencies but is a matter of absolute necessity.

Conclusion

A better grasp of the logical and hence necessary nature of the relation between evidence and hypothesis would have many benefits. It would give a sounder, more objective, philosophy of science and philosophy of the law of evidence. It would sort out the perennial puzzles of the correct interpretation(s) of probability. And, since evidence for hypotheses is evaluated by everyone every day, it would give an easily appreciated field where everyone understands necessities.

Notes

1 Introductions in Ian Hacking, *The Emergence of Probability* (2nd edn, Cambridge: Cambridge University Press, 2006), ch 2; D.C. Stove, *Probability and Hume's Inductive Scepticism* (Oxford: Clarendon, 1973), ch 1; D.H. Mellor, *Probability: A Philosophical Introduction* (London: Routledge, 2005), ch. 1.

2 Daniel Kahneman, Paul Slovic and Amos Tversky, *Judgement Under Uncertainty: Heuristics and Biases* (Cambridge: Cambridge University Press, 1982); general discussion in Nevin Climenhaga, 'Epistemic Probabilities are Degrees of Support, not Degrees of (Rational) Belief', *Philosophy and Phenomenological Research* 108 (2024): 153–76.

3 James Franklin, *What Science Knows: And How It Knows It* (New York: Encounter Books, 2009), ch. 3; James Franklin, 'Thomas Kuhn's Irrationalism', *New Criterion* 18, no. 10 (June 2000): 29–34.

4 Reviews in Michael Strevens, 'A Closer Look at the New Principle', *British Journal for the Philosophy of Science* 46 (1995): 545–61; Barry Loewer, 'David Lewis's Humean Theory of Objective Chance', *Philosophy of Science* 71 (2004): 1115–25.

5 James Franklin, *The Science of Conjecture: Evidence and Probability before Pascal* (Baltimore: Johns Hopkins University Press, 2001/2015), chs 1–3.

6 (Australian) *Evidence Act 1995* (Cth) s 135.

7 Ibid, s 137.

8 Stephanie Denison, Christie Reed and Fei Xu, 'The Emergence of Probabilistic Reasoning in Very Young Infants: Evidence from 4.5- and 6-month-olds', *Developmental Psychology* 49 (2013): 243–49; similar in Fei Xu and Vashti Garcia, 'Intuitive Statistics by 8-month-old Infants', *Proceedings of the National Academy of Science of the U.S.A.* 105 (2008): 5012–15; Chris A. Lawson and David H. Rakison, 'Expectations about Single Event Probabilities in the First Year of Life: The Influence of Perceptual and Statistical Information', *Infancy* 18 (2013): 961–82.

9 Erno Téglás, Edward Vul, Vittorio Girotto, Michel Gonzalez, Joshua B. Tenenbaum and Luca L. Bonatti, 'Pure Reasoning in 12-month-old Infants as Probabilistic Inference', *Science* 332 (2011): 1054–59.

10 David Stove, 'Deductivism', *Australasian Journal of Philosophy* 48 (1970): 76–98.

11 David Stove, *Popper and After: Four Modern Irrationalists* (Oxford: Pergamon Press, 1982).

12 E.g. E. T. Jaynes, *Probability Theory: The Logic of Science* (Cambridge: Cambridge University Press, 2003), Franklin, *What Science Knows*; Jan Sprenger and Stephan Hartmann, *Bayesian Philosophy of Science* (Oxford: Oxford University Press, 2019).

13 George Polya, *Patterns of Plausible Inference* (2nd edn, Princeton: Princeton University Press, 1968), 4. The derivation of it from Bayes' Theorem is as follows. Let h stand for hypothesis, e for evidence. Then by Bayes' Theorem, $P(h|e) = P(e|h) \times P(h) / P(e)$. Now if the evidence e is a consequence of the hypothesis h, then $P(e|h) = 1$. So, $P(h|e) = P(h) / P(e)$. If e is non-trivial, that is, not known with certainty already, then $P(e) < 1$. So, $P(h|e)$ equals $P(h)$ divided by a number less than 1. So, $P(h|e) > P(h)$, that is, the hypothesis is more probable on the evidence than it was before.

14 Daniel Kennefick, *No Shadow of a Doubt: The 1919 Eclipse That Confirmed Einstein's Theory of Relativity* (Princeton: Princeton University Press, 2019).

15 Example from Stephen F. Barker, *Induction and Hypothesis: A Study of the Logic of Confirmation* (Ithaca NY: Cornell University Press, 1957), 76.

16 Alan Hájek, 'The Reference Class Problem Is Your Problem Too', *Synthese* 156 (2007): 563–85; Mark Colyvan and Helen M. Regan, 'Legal Decisions and the Reference Class Problem', *International Journal of Evidence & Proof* 11 (2007), 274–85.

17 James Franklin, 'Feature Selection Methods for Solving the Reference Class Problem', *Columbia Law Review* Sidebar, Mar 2010; summary in 'The Objective Bayesian Conceptualisation of Proof and Reference Class Problems', *Sydney Law Review* 33 (2011): 545–61, section IX.

18 J. Maynard Keynes, *Treatise on Probability* (London: Macmillan, 1921).

19 James Franklin, 'Resurrecting Logical Probability', *Erkenntnis* 55 (2001): 277–305.

20 James Franklin, 'Case Comment: United States v. Copeland, 369 F. Supp. 2d 275 (E.D.N.Y. 2005): Quantification of the "Proof Beyond Reasonable Doubt" Standard', *Law, Probability and Risk* 5 (2006): 159–65.

21 Polya, *Patterns of Plausible Inference*, 8.

22 E. T. Jaynes, *Probability Theory: The Logic of Science* (Cambridge: Cambridge University Press, 2003).

23 In detail in Franklin, 'Resurrecting Logical Probability', sections 1–3.

24 James Franklin, 'Non-deductive logic in mathematics', *British Journal for the Philosophy of Science* 38 (1987): 1–18; many examples in Polya, *Patterns of Plausible Inference*.

25 Popular exposition in John Derbyshire, *Prime Obsession: Bernhard Riemann and the Greatest Unsolved Problem in Mathematics* (New York: Plume, 2004).

26 Donald Knuth, *The Art of Computer Programming* (3rd edn, Reading, Mass: Addison-Wesley, 1997), vol. 2 ch. 3; Antony Eagle, 'Chance Versus Randomness', *Stanford Encyclopedia of Philosophy*, 2010/2018.

27 D. C. Williams, *The Ground of Induction* (Cambridge, MA: Harvard University Press, 1947), ch. 4; Keith Campbell, James Franklin and Douglas Ehring, 'Donald Cary Williams', *Stanford Encyclopedia of Philosophy*, 2013/2023, section 6.

28 David C. Stove *The Rationality of Induction* (Oxford: Clarendon, 1986), ch . 6.

6

The grounding of logic in prelogical necessities

So far – especially in the two previous chapters on mathematical proof and logical probability – we have taken logic for granted. Usually, logic can be taken for granted in reasoning about any subject, but in an examination of necessities in general, there comes a time when we must turn our gaze on logic itself.

What is logic about? Logical necessity is as basic as necessities come. It has been a central object of study since Aristotle's work on the syllogism. But its nature remains unclear, and the question of its nature remains out of focus in philosophy.

It is normally thought that 'Logical consequence is the central concept in logic. The aim of logic is to clarify what follows from what, to determine which are the valid consequences of a given set of premisses or assumptions'.[1] Certainly, notions of contradiction, relevance, 'doesn't follow', proof, evidence for, proof beyond reasonable doubt, confirmation and refutation, dilemma and paradox are crucial to reasoning. Logic is thus essential to political debate, legal dispute, scientific theorizing, medical diagnosis and criminal investigation. All those are about how the facts and evidence *bear on* what is being claimed (that is, the logical relation of evidence to conclusion).

There certainly is such a field of study – the field of logical relations between propositions – and it is essential to understand it. But it does not follow that it is the most basic thing in logic, or the right place to start. Just as the ethics of right action is important, but dependent on a more basic ethical level of the worth of persons, or mathematics contains proofs but is about something more basic, namely quantity and structure, so there is a logic dealing with relations of propositions, but it depends on something more basic.

As so often in questions about what is philosophically fundamental, there are clues in Aristotle. Aristotle wrote a work on logical consequence, the *Prior Analytics*, but it is preceded in the canon of Aristotle's logical works by the *Categories*. That work is more a classification of things in the world 'spoken of' than about how to speak, think or reason about them,[2] but its contents are necessary for the logic that comes later. We might say it is about prelogic.

Thus, a sharp distinction is drawn between the (conceptual and/or linguistic) world of propositions, and the world that propositions are about or describe, the world of states of affairs, of how things actually are (though, of course, that does not preclude propositions themselves also being a 'world' that can itself be talked about). If logic is about how we (must) speak or think about the world, we need first to grasp what the world is like in order that that we can speak or think about it at all.

Prelogic and its necessities: The categorial structure of the world

Logic begins by assuming the subject-predicate structure of atomic propositions. 'Socrates is wise', 'Snow is white', 'Wisdom is more precious than rubies'; in general, 'Fa' and 'Rab'. Then there are logical combinations of those atoms: 'For all x Fx', 'If Socrates is a man and all men are mortal . . .' '$\exists y \forall x Rxy \rightarrow \forall x \exists y Rxy$'. Who said that the atoms of logical form are Fx and Rxy? What makes that so? Without answering that question, we have not started our investigation of logic at the most basic philosophical level.

It is only useful to speak of 'a is F' or 'x and y are R-related' if the linguistic composition of those sentences reflects the metaphysical composition of the states of affairs that are spoken of. The world is a world of states of affairs (facts, ways the world is), in the title of D.M. Armstrong's book.[3] 'The world is the totality of facts, not of things' – if Wittgenstein's gnomic dictum will bear that construction. As Aristotle says, 'Of things discoursed upon, some are enunciated after a complex, others after an incomplex, manner; the complex as "a man runs", "a man conquers", but the incomplex as "man", "ox", "runs".'[4] The world consists of things' having properties and relations, of Socrates' being wise, of snow's being white, of 3's being a divisor of 15. Logic is useful because that is how the world is, and we know that.[5]

That is not to say that the 'objects' that have properties need to be the 'medium-sized dry goods' of natural languages – tables being square, cats being furred, and so on. If the world is continuous stuff (as Aristotle thought), then it is inevitable that the fit is imperfect between discrete language-friendly concepts and the continuous world. Different natural languages may divide continua differently and have different borderline cases – as even French and English do with flowing water (river/stream/brook/creek do not correspond closely to French equivalents).[6] Developed science may regard stuffs, fields and wave functions as more the real fabric of physical reality rather than the macroscopic objects of ordinary experience.[7] But those 'objects' have the same metaphysical structure of properties and relations; for example, the wave function in quantum mechanics is a normal function on space-time that evolves according to a partial differential equation, describable by mathematics with standard logic. The basic metaphysics of science is still the same as it ever was: objects (in a wide sense) have properties and science studies those properties and their relations.[8]

If the world consisted only of angels, or (abstract) numbers, sense data or gas clouds with indeterminate boundaries, even those would have the same formal structure as logic requires. One might wonder, though, if there could be any kind of reality that is not like that. It is arguable that God may be too simple to be subject to universal-particular compositionality,[9] but that is rather beyond our ken, and any reality within the reach of our understanding seems likely to be as logic requires.

States of affairs are a type, so discussion of them includes possible states of affairs, those that could obtain but happen not to.[10] What 'possible' means needs some discussion, to take place below.

It is important to distinguish states of affairs (items in the world, or how things stand in the world) from propositions (in the conceptual or linguistic domain, whatever account is given of that, of which more below). The state of affairs of Socrates being wise obtains (or did obtain), while the proposition 'Socrates is wise' is true and the proposition 'Socrates is not wise' is false. There are (it is usually thought) no negative states of affairs corresponding to false propositions.

Thus, states of affairs, the furniture of the world, come first, prior to humans. The superstructure of propositions about states of affairs, true and false, comes later, when human thought and language occur. The structure of thought and language may be much more complicated than that of the states of affairs they are about (though the reverse is also true), and there could be mismatches between complex language and the world it purports to describe. The oldest Londoner is an item in the world, but the average Londoner is not; yet it is possible to understand what reality talk about the average Londoner describes. One can accept that 'for Socrates's death to take place just is for Socrates to die' despite the different grammatical forms of the two clauses.[11] The metaphysical structure of the world cannot be read off from language. For example, there do not have to be genuine disjunctive states of affairs corresponding to disjunctive propositions,[12] or anything in the world of states of affairs corresponding to 'if' (though there could be causal and other connections which make certain inferences justified). Those are matters for philosophical and scientific investigation.

What sorts of things are allowed for *a*, *F* and *R* in *Fa* or *Rab*? That is a difficult metaphysical question, addressed by Aristotle in the *Categories*. How much of that is essential for understanding logic and logical necessity is hard to say. We will consider the minimal amount as needed.

Before proceeding to the nature of propositions and the logical relations holding between them, it is desirable to consider what necessary relations, if any, hold between states of affairs themselves. Any such necessities would be very metaphysically basic, and it would be expected that logic cannot transgress them.

A crucial one is that the obtaining of a state of affairs excludes the obtaining of the contradictory state of affairs. As Aristotle puts it, 'the same attribute cannot at the same time belong and not belong to the same subject in the same respect'.[13] The belonging of attributes is a matter of states of affairs, not propositions. That has been taken to be a main reason for regarding (at least the most basic) logical truths as about the world rather than about thought or talk about the world.[14] Here we will prefer to say that such a truth is a prelogical or metaphysical truth, which logic (which concerns by definition propositions) must reflect: non-contradiction in propositions reflects non-contradiction in reality.

A generalization concerns the mutual exclusion of the determinates of a determinable. Colour is of its nature a determinable with a range of determinates, such as red and green, so that 'Nothing is both red and green all over at the same time' is a (prelogical) necessary truth.

It is usually thought by states-of-affairs theorists that there are conjunctive states of affairs, although there are no disjunctive ones.[15] In that case there are necessary connections between states of affairs, namely parthood: the state of affairs of this raven's being black is part of the state of affairs of these two ravens' being black. It is natural to think that that necessary connection between those states of affairs *grounds* the logical necessity of 'If these two ravens are black then this raven is black.' That simple example provides a paradigm for how logical truths (about propositions) can supervene on necessary relations between states of affairs.

In a previous chapter, we have seen examples of other kinds of necessary connections between states of affairs, such as that between being two rows of three crosses each and being three columns of two crosses each. So, the Humean thesis that 'there are no necessary connections between distinct existences' is not applicable to states of affairs in general, since states of affairs are not distinct in the sense needed to make the thesis true,[16] and there is no reason to believe that all states of affairs consist of atomic states of affairs that are so distinct. The possible distinct existences 'this page's being white' and 'this page's being black' do have a necessary connection, namely incompatibility.

Now that we have some grasp of 'prelogic', the basic metaphysical structure of reality that logic reflects, we are in a position to ask what the bearers of logical relations are, what the logical relations between them are, and how those relations connect with prelogic.

What are propositions?

It is normally, and reasonably, thought that logical necessity is a property of propositions (such as 'Snow is white or not white') or a relation between propositions (such as the relation of logical implication between 'All men are mortal' and 'Socrates, a man, is mortal'). So, the nature of propositions needs to be addressed first. It would be unsatisfactory to attempt an account of logical properties and logical relations without some account of the bearers of them.

Naively, a proposition can be thought of as what 'Snow is white' and 'La neige est blanche' have in common. Sentences in languages are said to be true or false but are so only in virtue of some proposition that they express. What kind of thing is a proposition, and prior to that, what kind of things are its constituents, the 'terms' 'snow/neige' and 'white/blanche'?

Twentieth-century Anglophone philosophy tended to avoid that question and to accept an implicit and unexamined 'Fregean-Platonist' answer. According to it, propositions (and their constituents) are eternal entities somewhat similar to numbers conceived Platonistically, existing in another realm – Frege's 'Third Realm' (neither physical nor mental).[17] Platonism about such entities is even more implausible than about numbers; all the usual arguments against Platonism apply and, in addition, the eternal existence of propositions about contingent human constructions is less reasonable than about numbers. '4 is twice 2' has an air of being an eternal verity, but surely not 'A majority of Americans prefer Doritos'.

It has not proved easy to find a convincing, more naturalist or Aristotelian, alternative. Here we adopt a 'causal-mentalist' theory of propositions, which integrates them with known human processes for forming knowledge.[18]

On this view, humans (and higher animals[19]) have built-in perceptual algorithms for clustering input and forming discrete internal symbols (or pre-symbols) that classify input. (A clustering algorithm is any recipe that takes in data points and 'sticks together' or divides into groups or 'clusters', those that are similar to one another.) For example, when seeing a cat against a background, the algorithm ties together colour-similar pixels of the object and those of the background and identifies the cat as a single object against that background. (The story here is thus a version of causal theories of reference.[20]) Vision is sensitive to features such as colour and texture, so an object, once identified, can be seen to have such perceptual properties that are shared by its pixels, such as being largely black. Thus, the outcome of the perceptual algorithm is a mental but pre-linguistic entity, with the *Fa*, subject-predicate structure that mimics the universal-particular structure of external states of affairs. That outcome is an atomic perceptual proposition. A true one, in the normal case, though perceptual algorithms like any others are not perfect.

While that proposition exists in a single mind, coordination across minds (needed to allow communication of propositions from one person to another) occurs because we all have (almost) the same perceptual algorithms. Just as your cake and mine resemble each other (have similar persistent properties) if we use the same recipe on similar ingredients, so your token internal proposition that there is a largely black-furred object before us resembles mine in virtue of our having run similar perception-processing clustering algorithms on similar visual input.

The mind – higher animal as well as human – is also equipped with another clustering algorithm that classifies the objects so recognized into kinds. It recognizes that small moving furred objects come in groups that have nearly common features within each group but without much in between, leading to internal classification into kinds that (in humans but not other higher animals) can later be given the labels 'cats' and 'dogs'.

Humans (but apparently not other animals) can acquire sufficient awareness of the compositionality of their internal symbols, '*F*' and '*a*', to be able to mix and match them. That is aided by the ability of humans to express their internal propositions in language, which requires a further step of clustering, which is a precondition of the learnability of language. For an infant to associate the sound 'cat' with experiences of cats, it must have solved four perceptual discretization or clustering problems first: (1) It has to cut individual cats out of the visual background, (2) it has to classify those objects as (potentially) a single kind, then (3) it has to segment the sound stream so that the syllable 'cat' is isolated, and finally, (4) it has to classify that occurrence of 'cat' with other occurrences of that syllable. Only after that can there be an association learned between the syllable-type 'cat' and the experience of the type cats. Thus, organizing perceptual experience into discrete repeatable pieces must take place before questions about the meaning of words can even arise.

There is thus a close relationship between propositions and possible states of affairs, though they are not the same. Typically, perceptual propositions are contingent, so that if 'the cat is black' is true, then 'the cat is not black' is false but names a possible state of affairs. A major reason for entertaining propositions includes 'mental time travel',[21] the considering of non-obtaining scenarios, often with a view to bringing them about or avoiding them. For that purpose, we need to begin with atomic perceptual propositions induced via ordinary perceptual processes by states of affairs that obtain, but then 'mix and match' using the recombination of the discrete parts of '*Fa*'. Analysis of the true statement 'John loves Mary' will lead to the recognition that 'Mary loves John' is a possible state of affairs (with no logical connection to the original).

There is not much point in entertaining the proposition that all round rocks are square, as we know it is impossible. But as our knowledge is limited, we do have to consider the propositions that the Riemann Hypothesis is true and that it is false, although one of those is impossible. Thus, there can, in principle, be propositions 'corresponding to' impossible states of affairs, but that is a special case and, in normal life, propositions are aligned with possible states of affairs.

Supervenience of logic on relations between states of affairs

To move beyond atomic perceptual propositions, the human mind (but the animal one only rudimentarily) has a range of logical and other epistemic abilities, from logically combining atomic propositions to inferring unseen entities (such as other minds), confecting fictions, and talking reflexively and explicitly about logic, language and fictions. The structure of the resulting thought and language may contain more than the world does (that is the point of fictions) but in many cases retains a close relation to states of affairs – in science especially one intends the complexities to stay anchored to what is out there.

Tarski's well-known work on how truth flows from atomic propositions through to logical combinations of them is one account of how logically complex propositions retain their connection to states of affairs. A simple example shows the nature of the connection. A particularly clear case is the close connection between the logical consequence relation between (certain) propositions and the inclusion relation between states of affairs, described above: The state of affairs of this raven's being black is a *part* of the state of affairs of this raven and its neighbour both being black. It is thus impossible for the latter state of affairs to obtain without the former also obtaining. The inclusion relation between the states of affairs is the truthmaker of the fact that it is impossible for the second proposition to be true without the first also being true. That is, the latter proposition implies the former.

Similarly, the truthmaker of the validity of the syllogism (read extensionally),

All As are Bs
All Bs are Cs
Therefore all As are Cs

is the (necessary) transitivity of the inclusion relation between sets. Propositions are (types of) mental entities, while sets are (on an Aristotelian, non-Platonist view) in the world, and the necessary properties of the inclusion relation ground the necessity of the logical consequence relation between the premises of the syllogism and the conclusion.

It is possible to answer similarly the question that Penelope Maddy asks about the disjunctive syllogism:

> What's hidden in my hand is either an ordinary dime or a foreign coin of a type I've never seen. (I drew it blindfolded from a bin containing just these two types of objects.) It's not a dime. (I can tell by the feel of it.) Then, obviously, it *must* be a foreign coin! But what makes this so?[22]

What makes it so – its ground in extra-logical reality – is the incompatibility of properties (local or foreign, whose definitions make them stand in the same relation as red and not red). If we were to generalize and say that any disjunctive syllogism is valid: If p-or-q is true and p is false then q is true, that is similarly grounded – in the incompatibility of true and false.

However, the fact that thought and language can be more complex than the world they describe creates complications in the relation of the truth of propositions to the obtaining of states of affairs. Medieval logic had a sound sense of that with its theory of 'second intentions'. For example, in saying that 'Virtue is a genus', or 'these propositions are contradictories', it is not implied that genus and contradiction are among worldly properties – they are modes of organizing thinking about the world.[23] Likewise, negations and privations such as holes are 'mental beings with foundation in the real': ways to think correctly about reality but not subject to fundamentalist interpretations as items in reality.

In light of that, we are freed to admit that *some* logical truths are just about thought or language. It is natural to say that the logical necessity of 'If p, then p&p' does not rest on any deep necessities about states of affairs but simply on the fact that one has repeated oneself. Individual states of affairs cannot repeat or clone, but propositions and sentences can. The proposition 'p&p' therefore does not identify any state of affairs other than that identified by 'p', which is a fact about thought and language rather than the world.

So, it is perfectly possible that there are such things as trivial logical necessities, arising from the complexities of thought and language. Such examples should not blind us to the more significant cases where logical necessity is anchored in the world.

Logic as truth in all possible worlds/models

We are now in a position to support the most promising of the several theories about the nature of logical truth and logical consequence, the Tarskian one that it is truth in all models. We will see though that it is not a totally self-contained answer and makes assumptions about prelogical necessities 'out there'.

Tarski writes that 'We say that the sentence X follows logically from the sentences of the class K if and only if every model of the class K is at the same time a model of the sentence X.' Gila Sher writes, 'A sentence S is a logical consequence of a set of sentences C iff there is no model in which all the sentences of C are true and S is false.'

The theory does have the advantage of connecting logical validity with truth 'out there': It is 'semantic' in Tarski's sense of 'relations between the expressions of language and the objects about which these expressions speak'[24] (as opposed to semantics in the sense of the meaning or words or concepts). It thus gives the opportunity to connect logic with the prelogical structure of the world described above. As Sher concludes:

> For a logical system to be adequate, it cannot say that S follows logically from S1 ,. . ., Sn unless there is an appropriate connection between the worldly conditions that would render all of S1, . . ., Sn true and the worldly conditions that would render S true.[25]

It is very difficult to make this idea work, as a complete definition of logicality. As thus stated, the idea has a number of holes:

1. What counts as a 'model' (of a set of sentences)? It involves allowing the meaning of the nonlogical terms in the sentences to vary but not the logical ones (for example in 'All men are mortal' one can vary the interpretation of 'men' and 'mortal' but not of 'all').[26] As Tarski himself says, one must have the division into logical and non-logical constants first. He says this is not clear and he is doubtful if it can be done.[27]
2. Secondly, 'every model' means 'every possible model'. Are we sure we understand 'possible' in 'possible worlds' or models? If it means 'logically possible', that would be circular, if our aim is to explain logical necessity without remainder.
3. The theory does not distinguish between logical necessity and other absolute kinds of necessity, such as metaphysical and mathematical necessity. In all models in which a sentence is true, all metaphysical and mathematical consequences as well as all logical consequences of it will be true. For example, the axiom of infinity is true in all models (if it is, as it seems, a necessary truth about numbers), but it is agreed that it is not logical – that was one of the main sticking points of Russell's logicist view of mathematics.
4. The theory rules out non-deductive logic *a priori* (since what is probabilistically confirmed by a set of sentences is not true in *all* models).

So, the Tarskian theory of logic has strong intuitive appeal but is at best incomplete without further clarification.

The holes would be plugged if there were a separate criterion for distinguishing between the logical and the non-logical. It would distinguish between non-logical and logical language, and hence between logical and other kinds of necessity and it would, in principle, be applicable to non-deductive logic.

This task is attempted, with considerable success, in the work on 'formal' necessity by Penelope Maddy and especially by Gila Sher's theory of invariance. We notice that:

> 'All men are mortal and Socrates is a man, therefore Socrates is mortal.'

is valid for the same reason as:

> 'All men are mortal and Plato is a man, therefore Plato is mortal.'

So, we conclude that logical consequence ought to obey the symmetry of individual constants. That is, we can substitute any individual for 'Socrates'; the validity of the inference is invariant under substitution of individuals. So, the necessity of the inference belongs to the formal pattern:

'All men are mortal and X is a man, therefore X is mortal.'

'X' may be allowed to range over any category of individuals, including for example numbers and logical constants. (It is initially surprising and must have surprised the Greeks who first discovered logic, that there are inference patterns that apply across all categories of things whatsoever, but that is how it turned out.) As Tarski puts it, logical consequence 'cannot be affected by replacing the designations of the objects referred to in these sentences by the designations of any other objects'.[28] One can say that a model of a sentence is a possible world in which different individuals replace the ones in the actual world (this is intended as a necessary condition for logicality, not a sufficient one).

But that must mean that possible individuals ('actual-counterfactual individuals'[29]) are included in models – if in the actual world 'chordate' and 'cordate' are co-extensive, that does not make the two logically equivalent; there are possible though no actual individual animals that are one but not the other. But then, the question is again raised, what kind of possibility? And why not models with impossible individuals, such as square circles and non-self-identical objects? Why are those not allowed in models? It seems that models will have to be populated with all and only logically possible individuals, which is feasible but means that the problem of logicality has not yet been solved.[30] One is relying on necessity in some extra-logical sense, in the real world with its prelogical necessities. So, the theory that logical truth is truth in all models is correct but needs some external theory of the prelogical and kinds of necessity to specify what counts as a possible model.

The wider sphere of topic-neutral necessities

The story of absolute necessities would not be complete without noticing that beyond the sphere of strictly logical necessity there is a penumbra of a priori and extremely general arguments that work in all possible worlds – extrapolation, sorites, confirmation of theories by consequences, induction, symmetry arguments, set theoretical arguments. In virtue of what are they reasonable? For example, is this argument logic?

Most As are Bs
Most As are Cs
Therefore some Bs are Cs.[31]

Certainly, it is impossible for the premises to be true and the conclusion false. It superficially resembles standard syllogistic arguments. It appears to be as general and topic-neutral (that is, not specific to any type of subject-matter) as they are. But to explain why it works, one must talk about counting in sets: because two majorities in a set must overlap, there must be an element in common among the Bs and the Cs. Does that make it mathematical rather than logical? There seems to be no principled division between the mathematical and the logical in such cases.

The porosity of division between mathematics and logic is also revealed by examples like:

$1 + 1 = 2$

(For any *F* and *G*), if there is an *F* and a *G* and no *F*s are *G*s then there exist *x* and *y* such that *x* and *y* are *F* or *G* and $x \neq y$.

The first is clearly mathematical and the second clearly logical, but they are equivalent – perhaps not logically equivalent, but with an intimate necessary connection.[32] Given that mathematics is the science of structure and hence of complexity, as soon as logic becomes complex it will incorporate mathematical truths.

We considered in Chapter 5 one of the most central non-deductive inference principles, the proportional syllogism. A typical example is:

The great majority of As are Bs
This is an A
Therefore very probably this is a B.

Again, the argument appears to be as general and topic neutral as the syllogism, and the fact that the conclusion follows with less than certainty from the premises seems insufficient reason to deny it is logic. If one were to decree that logic was only deductive, that is, about arguments whose conclusions followed with certainty, one would simply be left with the need to give an account of arguments which resembled logical ones in all other respects than following with certainty. The premise-conclusion nexus of 'partial implication', in Keynes' words, is *prima facie* of the same nature as implication.

Similarly, with mereological arguments.

A is part of B
B is part of C
Therefore A is part of C

is just the syllogism with 'subset of' replaced by 'part of' so it appears to be as general, necessary and topic neutral.

Then there is a swathe of arguments whose strength depends on continuous variation.[33] They include slippery slope arguments, extrapolation arguments, sorites, fuzzy logic and vagueness. Similarity is as general and topic-neutral as identity, and where identity is transitive, similarity is near-transitive, as in:

> Crimson is very like scarlet.
> Scarlet is very like vermilion.
> Therefore, crimson is somewhat like vermilion.

An example of an extrapolation argument is D. M. Armstrong's 'argument from microscopes' for the reality of the sub microscopic: We tend to believe in the real existence of new things we see through magnifying glasses, because of the continuity with what we see with the naked eye. D. M. Armstrong argues:

> We would trust these new deliverances because we had been able to check the native eye to some extent. We could then substitute a more powerful glass, checking its reliability by reference back to the original glass; and so proceed by easy stages to the most powerful microscope.[34]

Similar arguments are familiar in calibrating methods of ancient dating and measuring the distance of galaxies.

We are getting close to some of the more general necessities involving structured spaces of universals, such as the betweenness relations between colours and the transitivity of the 'greater-than' relation between quantities such as lengths. Those sorts of necessities can ground arguments that have relation to quantities.

Again, one concludes that the number of known absolute necessities is much larger than normally acknowledged.

Conclusion

It is widely agreed that logic is full of necessities. But views on the nature of logical necessity are very diverse and have been hampered by thinking of logic as solely about concepts and words. What has not been understood is that the necessities of logic reflect necessities, here called 'prelogical', in the world. The universal-particular structure that is inherent in all (or almost all) reality is essential to the ability of the subject-predicate structure of atomic propositions to enable thought and communication.

For all the most general forms of reasoning, the structure of the world's necessities must cooperate with the forms of human reasoning to make those forms valid and enable us to infer new truths from old. The world does that.

Notes

1 Stephen Read, *Thinking About Logic: An Introduction to the Philosophy of Logic* (Oxford: Oxford University Press, 1995), 35.

2 Paul Studtmann, 'Aristotle's *Categories*', *Stanford Encyclopedia of Philosophy*, 2021.

3 D. M. Armstrong, *A World of States of Affairs* (Cambridge: Cambridge University Press, 1997); other possible views on the fundamental structure of reality and their relation to logic discussed in Michaela Markham McSweeney, 'Logical realism and the metaphysics of logic', *Philosophy Compass* 14 (2019): e12563.

4 Aristotle, *Categories*, ch. 2.

5 A position called 'metaphysicalism' in Agustín Rayo, *The Construction of Logical Space* (Oxford: Oxford University Press, 2013), part 1.2.

6 Helen Bromhead, *Landscape and Culture – Cross-linguistic Perspectives* (John Benjamins, Amsterdam, 2018), ch. 2.

7 Emphasized in James Ladyman and Don Ross, with David Spurrett and John Collier, *Every Thing Must Go: Metaphysics Naturalized* (Oxford University Press, Oxford, 2007); other aspects of the match and mismatch between language and world in Brian Cantwell Smith, *On the Origin of Objects* (Cambridge, MA: MIT Press, 1996).

8 James Franklin, *What Science Knows: And How It Knows It* (New York: Encounter Books, 2009), ch. 4.

9 William F. Vallicella, 'Divine Simplicity', *Stanford Encyclopedia of Philosophy*, 2006/2023.

10 On a matter of terminology, Armstrong uses 'states of affairs' to mean 'facts', that is, states of affairs that actually obtain, whereas 'states of affairs' is more naturally used of the type, so can include possible states of affairs. This distinction will not be important in what follows.

11 Contrary to Rayo, *The Construction of Logical Space*, part 1.2.

12 Armstrong, *World of States of Affairs*, 27, 173.

13 Aristotle, *Metaphysics* 1005b19-20.

14 Tuomas E. Tahko, 'The Law of Non-Contradiction as a Metaphysical Principle', *Australasian Journal of Logic* 7 (2009), 32–47; further in Tuomas E. Tahko, 'The Metaphysical Interpretation of Logical Truth', in Penelope Rush ed, *The Metaphysics of Logic* (Cambridge University Press, Cambridge, 2014), 233–48; a treatment of the epistemology in Penelope Rush, 'Logical Realism', in Rush, *Metaphysics of Logic*, 13–31.

15 Armstrong, *World of States of Affairs*, 19, 35.
16 Catherine Legg and James Franklin, 'Perceiving Necessity', *Pacific Philosophical Quarterly* 98 (2017): 320–43, section 2.
17 Michael Dummett, 'Frege's Myth of the Third Realm', in Michael Dummett, *Frege and Other Philosophers* (Oxford: Oxford University Press, 1991), 249–62; Gustavo Picazo, 'The Long Shadow of Semantic Platonism', *Philosophia* 49 (2021): 1427–53.
18 Summarized from Jeremiah Joven Joaquin and James Franklin, 'A Causal-Mentalist View of Propositions', *Organon F* 29 (2022): 47–77.
19 The abilities of higher animals in forming discrete representations and reasoning with them surveyed in Michael Rescorla, 'Chrysippus's Dog as a Case Study in Non-Linguistic Cognition', in *The Philosophy of Animal Minds*, ed. R. W. Lurz (Cambridge: Cambridge University Press, 2009), 52–71.
20 As in Michael Devitt and Kim Sterelny. *Language and Reality: An Introduction to the Philosophy of Language*, 2nd edn (Boston, MA: MIT Press, 1999).
21 Thomas Suddendorf and Michael C. Corballis. 'The Evolution of Foresight: What Is Mental Time Travel, and Is It Unique to Humans?' *Behavioral and Brain Sciences* 30, no. 3 (2007): 299–313.
22 Penelope Maddy, 'A Second Philosophy of Logic', in Rush, *Metaphysics of Logic*, 93; further in Penelope Maddy, *Second Philosophy: A Naturalistic Method* (Oxford University Press, Oxford, 2007), Part III.3; the ability of dogs in the disjunctive syllogism in Ulrike Aust, Friederike Range, Michael Steurer and Ludwig Huber, 'Inferential Reasoning by Exclusion in Pigeons, Dogs, and Humans', *Animal Cognition* 11 (2008), 587–97.
23 Matthew K. Minerd, 'Thomism and the Formal Object of Logic', *American Catholic Philosophical Quarterly* 93, no. 3 (2019): 411–44.
24 Alfred Tarski, 'The Concept of Truth in Formalized Languages (1933), in *Logic, Semantics, Metamathematics*, ed. J. Corcoran (Indianapolis: Hackett, 1983), 152–278, at 252; Gila Sher, *Logical Consequence* (Cambridge: Cambridge University Press, 2022), 6.
25 Gila Sher, 'Logical Realism – A Tale of Two Theories', in *The Philosophy of Penelope Maddy*, ed. S. Arbeiter and J. Kennedy (Cham: Springer, 2024), 241–61.
26 E.g. Sher, *Logical Consequence*, 14.
27 Alfred Tarski, 'On the Concept of Logical Consequence' (1936), in *Logic, Semantics, Metamathematics* ed. Corcoran, 409–20, at 418–19, quoted in Sher, *Logical Consequence*, 23.
28 Tarski, 'On the Concept of Logical Consequence', 415; Sher, *Logical Consequence*, 9.
29 Sher, *Logical Consequence*, 32.
30 Similar objections made to David Lewis's possible worlds theory in Scott A. Shalkowski, 'Modalism', in *The Routledge Handbook of Modality*, ed. Otávio Bueno and Scott A. Shalkowski (London and New York: Routledge, 2021), ch. 10.

31 An example considered by Lambert in the eighteenth century, James Franklin, 'Artifice and the Natural World: Mathematics, Logic, Technology', in *Cambridge History of Eighteenth-Century Philosophy*, ed. Knut Haakonssen (Cambridge: Cambridge University Press, 2006), 817–53, at 842; discussion in Sher, *Logical Consequence*, 25.

32 Discussion in Owen Griffiths and A. C. Paseau, 'Isomorphism Invariance and Overgeneration', *Bulletin of Symbolic Logic* 22 (2016): 482–502.

33 James Franklin, 'Arguments Whose Strength Depends on Continuous Variation', *Informal Logic* 33 (2013): 33–56.

34 D. M. Armstrong, *Perception and the Physical World* (London: Routledge, 1961), 159–60.

Part Four

Absolute Necessities Constrain Even God

7

Two great themes of mathematics: discrete/continuous and local/global

After the excursion in the last three chapters into matters connected with logic, we return in this and the next chapter to mathematics proper. We expand the range of mathematical necessities discoverable in ordinary reality by examining two overarching themes in mathematics, which identify commonalities among a number of disparate 'applications' of mathematics.

Global versus local, discrete versus continuous, simple versus complex, linear versus nonlinear, deterministic versus stochastic, ordered versus random, analytic versus numerical, constructive versus nonconstructive – those contrasts are among the great organizing themes of mathematics. They are forks in the road of mathematical technique – the concepts along one direction from the fork are very different from those along the other, even when they give complementary views on the same phenomena. Well-known though they are to even beginners in mathematics, it is hard to find a clear exposition of any one of those dichotomies. We consider just the first two.

Discrete mathematics (arithmetic, algebra, combinatorics, logic) versus continuous (geometry, calculus)

Perhaps the most deep-rooted contrast in mathematics is that between discrete and continuous. It is so ubiquitous in mathematics that the lack of a straightforward overview of the whole topic and explanation of its significance is astonishing.[1] Perhaps that is partly to be explained by the demise after the eighteenth century of the old Aristotelian philosophy of mathematics, according to which mathematics is the science of quantity, and quantity is divided into the discrete (studied by arithmetic) and the continuous (studied by geometry). But that seems an insufficient explanation, as the dichotomy remains fundamental to mathematics.

The basic distinction is clear enough, but hard to define in terms of anything simpler. **Discrete:** 'separate; detached from others; individually distinct'; **Continuous:** 'extending without interruption of substance; having no interstices or breaks; having its parts in immediate connection' (*Oxford English Dictionary*). The integers are discrete, the real line is a continuum. Matter may be continuous or discrete (atomic) – it cannot be determined a priori and scientific investigation is needed. Space and time appear to be continuous, but it is not impossible that they are discrete at a very fine scale.

The Greeks: the incommensurability of the diagonal and harmonies in music

The significance of the discrete/continuous distinction, which established it as one of the great themes of mathematics, became clear with the ancient Greek discovery of the incommensurability of the diagonal. It is necessary to explain this discovery in its own terms, as the common modern reinterpretation of the result as the 'irrationality of the square root of 2" obscures its original meaning.

In the most obvious continuous cases, like length, one chooses a unit arbitrarily and measures the ratio of all other lengths to the unit. That can give the impression that continuous quantity is not fundamentally different from discrete: to convert any continuous problem to a discrete one, it is just necessary to find a small enough unit to measure all the continuous quantities involved. Given a ruler divided finely enough, it should be possible to find the ratio of any continuous quantities, say lengths, by counting exactly how many times the ruler's unit is needed to measure each quantity. One length might be 127 times the unit and another 41 times, showing that the ratio of the lengths is 127 to 41.

That natural and even compelling thought is incorrect, as the ancient Greeks discovered. Perhaps the first truly surprising result in mathematics was the one attributed (traditionally but without much evidence) to Pythagoras, the proof of the incommensurability of the side and diagonal of a square. There is *no* unit, however small, which measures diagonal and side, that is, of which both are whole number multiples. (The method by which the Pythagoreans discovered this is unknown. No method is entirely easy.[2])

So, the ratios of continuous quantities are more varied than the relations of discrete quantities. The Greeks drew the conclusion, surely rightly, that geometry, and continuous quantity in general, is in some fundamental sense richer than arithmetic and not reducible to it via choice of units. While much about the continuous can be captured through discrete approximations, it always has secrets in reserve.

The second major discovery of the Greeks was that the continuous sometimes naturally gives rise to discrete structures, structures not immediately visible but which give a deep insight into the original continuous phenomenon. The original discovery of this kind, also attributed to Pythagoras, was the connection between the harmony of notes and the whole-number ratios of the lengths of the strings producing them. Although the lengths of strings and the pitch of notes are both quantities whose nature is to vary continuously, a discrete pattern emerges from them: for strings of a fixed tension, the easily perceived harmonies of octave, fifth and so on are produced by pairs of strings whose lengths are in small whole-number ratios. The simplest non-trivial ratio, 2:1, produces the most prominent harmony, the octave. Later science has explained the mystery and at the same time found one mechanism for how continuous variation can produce discreteness: pitch is caused by the frequency of soundwaves, and there is something special about the frequencies of waves in small integer ratios. They interact to produce regular large peaks that affect the eardrums, which randomly attuned frequencies do not.

Since then, the interplay between the discrete and the continuous has been at the heart of some of the most profound advances in science and mathematics. Examples include the eigenvalues of linear maps, resonance, algebraic topology and the way atoms and molecules arise as discrete solutions of the Schrödinger Equation (a partial differential equation in continuous space and time).[3] Another front in the relation between the two has been the extensive research in numerical analysis needed to enable (discrete) computer simulations to reliably model continuous processes such as fluid flow.

Discreteness in science: atomism and Mendelian genetics

We now turn to some more applied fields and ask what evidence is relevant to deciding whether a reality being studied is discrete or continuous. If something is continuous as far as perception can tell, it can be very hard to know if it is discrete at a smaller scale. But sometimes a signature of sub-microscopic discreteness is discoverable at a larger scale.

What is the evidence that atoms exist? The ancient Epicureans and some seventeenth-century philosophers proposed them as neat explanations of various phenomena, but they could not point to any convincing cases where an explanation in terms of atoms was superior to one using the alternative and more natural hypothesis that matter is continuous. The first convincing argument for atoms was that of Dalton, who discovered instances of what is now called the law of multiple proportions – if two elements form more than one compound between them, then the ratios of the masses of the second element which combine with a fixed mass of the first element will be ratios of small whole numbers. An atomic structure of matter gives a good explanation while a continuous one has little prospect. Dalton wrote:

> The element of oxygen may combine with a certain portion of nitrous gas, or with twice that portion, but with no intermediate quantity. In the former case nitric acid is the result; in the latter nitrous acid . . . Nitrous oxide is composed of two particles of azote and one of oxygen. This was one of my earliest atoms. I determined it in 1803, after long and patient consideration and reasoning.[4]

Another easily appreciated example is Mendel's work on genetics. One of the many problems with Darwin's original theory of evolution concerned 'blending inheritance'. Natural selection, according to Darwin, acts on small mutations that occur rarely and at random in populations. But since the possessor of such a mutation must breed with the 'normals' of the population, surely a favourable mutation should blend back into the population over a few generations before natural selection has a chance to act? Like many other problems with evolutionary theory, it was only admitted that it was serious once it had been solved. The solution came with Gregor Mendel's discovery that the genetic material was, often at least, discrete. And the discovery of discreteness, through Mendel's experiments with peas in his monastery garden, came, as it had with the Pythagoreans and Dalton, with the noticing of small whole-number ratios in material that was continuous in its nature.

Mendel took two lines of peas, short and long, each bred over several generations to be 'pure' or 'true'. That is, the short ones produced only short ones and the tall ones only tall ones. Then he crossed them, that is, pollinated the flowers of the short ones with pollen from the tall ones. With a genius for 'population thinking' exceedingly rare in his age, he thought to perform this experiment many times and count the results. He found, not very excitingly, that all the crosses were tall (not somewhere between short and tall, but clearly tall). But if this generation of tall ones were crossed with one another, the results were not all tall but consistently about three-quarters tall and one-quarter short. The simple phrase 'consistently about three-quarters . . .' is deceptive. It makes something hard to find look easy. The discreteness of the ratio 3:1 does not reveal itself in any single experiment, especially a small one with one or two plants. One must take a count in a large number of experiments and be content that the ratio is not exactly 3:1.

Mendel correctly posited a theory of discrete factors or genes that would explain the simple ratio. He supposed that length of peas depends on each individual's having two genes for height, each either 'tall' or 'short', with the combination 'tall' and 'short' resulting in the same appearance as two 'talls', and with random assortment of genes between generations. When first-generation plants (all appearing tall but in fact all having one 'tall' and one 'short' gene) are crossed with one another, the next generation has about one-quarter 'tall-tall', half 'tall-short' and a quarter 'short-short'; only the last quarter appear short.

With one bound, Darwin's theory was free of the blending inheritance problem. Assortment of discrete genetic material is capable of giving a mutation several chances to express itself fully.

It remains a mystery, however, whether space and time are discrete. The standard modern physical theories, relativity and quantum mechanics, are expressed in terms of continuous space and time but there are no observations – perhaps no possible observations – to confirm this directly. Erwin Schrödinger, like many physicists dealing with the very small, was impressed with how elaborate the structure of the mathematical continuum is and how little observational support there is for supposing it is instantiated in its entirety in real space.[5] But despite many proposals for having apparently continuous space and time arise out of something simpler, possibly discrete, none have become established.[6] Nor have such approaches been ruled out.

It is a contingent matter, as far as we know, whether space, time, matter, genetics or musical sounds have a discrete or continuous structure. But the structure they have then determines what kind of mathematics is applicable to them, that is, what necessities they instantiate.

Local versus global: in topology, differential equations, functional analysis

The distinction between local and global – between what is possible on a small scale but impossible when everything is fitted together on a large scale – is another of the great themes of mathematics. Like the discrete/continuous distinction, it is 'well-known' and used unreflectively across a range of areas, but is not explained in general.[7] Just as the lack of work on the discrete/continuous distinction may be partly attributed to the philosophical neglect of mathematics as the science of quantity, so the lack of attention to the local/global distinction may be partly attributable to the failure to see mathematics as the science of structure. A failure to see mathematics as being about real aspects of the physical world will leave no space to discuss those particular mathematical aspects.

The flavour of mathematical results about local-global interactions is well displayed by Euler's classic eighteenth-century paper on the bridges of Königsberg, already discussed in Chapters 2 and 4 above (see Figure 2.2). The citizens of Königsberg noticed that it seemed to be impossible to walk across all seven bridges over the River Pregel, without walking across at least one of them twice. Euler proved mathematically that they were right: although it is easy at any point to choose a bridge to walk over, and also to choose small portions of the figure and walk over all the bridges in them once only, it is absolutely impossible to perform the task as a whole. Although God could make bridges, islands or citizens differently, he could not make them as they are while at the same time

Figure 7.1 Escher's Waterfall.[8]

making it possible for the citizens to walk over all the bridges once and once only. What is easy locally is, provably, impossible globally — and with an impossibility as strong as impossibilities come, a mathematical impossibility.

Another classic example is the impossibility of building a circular staircase that goes up all the way round and ends at its starting point.[9] Many people are familiar with the famous Escher drawings which seem to show this kind of thing happening (see Figure 7.1), but are not explicitly aware that what makes the image so visually arresting is its apparent depiction of a mathematical impossibility.

The essential structure of the example is more clearly illustrated in Figure 7.2 below:

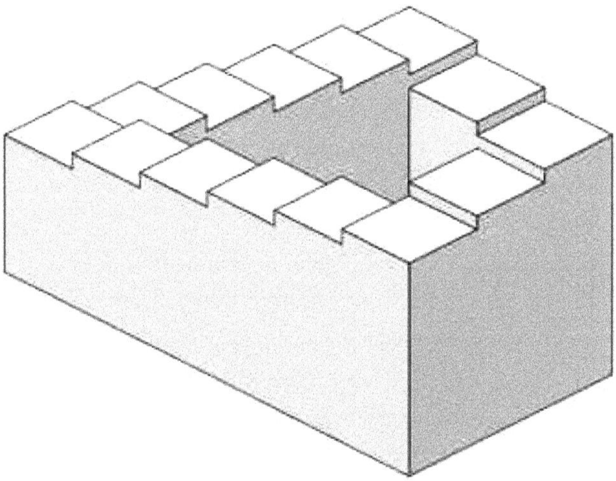

Figure 7.2 Endless stairs.[10]

It is possible to realize in three-dimensional space any small portion of the staircase, but not the whole staircase. Our imagination easily appreciates the local possibility of constructing a small piece of rising staircase but has more trouble with the global impossibility (although in this case the global impossibility is still simple enough to grasp more or less completely). The impossibility is not just empirical, since no change in the laws of nature would make such a staircase possible. A purely mathematical fact underlies the impossibility, namely, that there exists no continuous function from the circle to the real numbers which is increasing all the way round. The proposition therefore has nothing to do with the Euclideanness or otherwise of space or its global topology; in any substantial portion of a space where there is a coherent notion of 'up', the statement is true (as 'up' is a continuous function from the space to the real numbers, to which the theorem applies).

There are many more advanced theorems in some of the central fields of modern pure mathematics, like algebraic topology and differential geometry, of a similar character to the staircase example. They describe how global structure constrains local structure (or, depending on one's point of view, how local structure gives rise to global). A well-known instance is the 'hairy ball theorem', which states that the hair on a ball cannot be combed flat everywhere (formally, every continuous tangent vector field on the sphere vanishes somewhere).[11] The fact that this is impossible on the sphere while possible on the torus indicates that the theorem is about 'relationships between the local differential properties of a space and its topologic structure as a whole'[12] – how the global topology of the space constrains what is possible differentially locally.

A last example from mathematics is the four-colour theorem. One of the most famous, and difficult, theorems in mathematics states that any map (on a plane) can be coloured so that adjoining regions have different colours, using only four colours.[13]

These mathematical results, and the four-colour theorem in particular, bear out Thomas Reid's claim, 'Mathematics afford many instances of impossibilities in the nature of things, which no man would have believed if they had not been strictly demonstrated. Perhaps, if we were able to reason demonstratively in other subjects, to as great an extent as in mathematics, we might find many things to be impossible, which we conclude without hesitation, to be possible.'[14]

It is true that some popular philosophies of mathematics, such as Platonism, imply a divorce between pure mathematics and the physical world which would mean that the provable necessities of pure mathematics could not apply directly to the physical world. Those philosophies have difficulty making sense of examples such as those above. As explained in Chapter 3, an Aristotelian philosophy of mathematics does not have that problem, as it allows mathematical properties and their necessary interrelations to be instantiated in the physical world.

Local and global in other disciplines

Actually, there are several disciplines outside mathematics where necessities of the same kind are known, although it is not always appreciated that they are strictly mathematical. Another example of the difficulties of achieving a global best result, in truth mathematical but in an application well known to the humanities, is tuning and temperament in music.

As soon as harmony – the playing of two notes at once – is introduced into music, the problem arises about what choice should be made for the pitches on instruments such as pianos of the notes between one octave and the next. As described above, notes sound harmonious or consonant if there is a simple integer relation between the frequencies of their sound waves. Thus, two notes an octave apart have a ratio of 2:1; the next simplest ratio is 3:2, which is called the fifth and is the next most consonant in sound. To the ear, a very slight mistuning is acceptable, but ratios as close to 1 as 81:80 sound severely dissonant. If one tries to fill in the octave with a small number of notes (such as twelve, the standard number of notes between octaves on a piano), with allowance made for the fifth, the major third (5:4), and the major sixth (5:3), it turns out to be mathematically impossible to have them all exactly tuned. Some compromise must be made. The usual modern choice has been the system of "equal temperament", which makes each pair of consecutive notes on the piano differ by an equal amount, one-twelfth of an octave (that is, the ratio of their frequencies is always $\sqrt[12]{2}$:1). This means that none of the main ratios between notes is exactly correct, except for the octave itself. The major sixth in particular is substantially too sharp, a problem particularly evident on instruments of purer tone like the organ. Some compromises are better than others, but it is impossible to avoid making some compromise.[15]

In economics, Adam Smith's 'invisible hand' is a metaphor for the difference between what individual actors intend locally and the global effect of their actions. A buyer or seller 'generally, indeed, neither intends to promote the public interest, nor knows how much he is promoting it . . . . he intends only his own security; and by directing that industry in such a manner as its produce may be of the greatest value, he intends only his own gain, and he is in this, as in many other cases, led by an invisible hand to promote an end which was no part of his intention'.[16] The 'invisible hand' is not an entity or in any way a cause that acts. All the causes acting are local: the individual decisions of buyers and sellers. The global effect, overall prosperity, is the working out of the sum total of those local effects, in the same way as the solution of a differential equation is the sum of the effects of local actions. The sum of local effects necessitates the global effect, whatever that may be.

A modern example is the well-known Arrow Impossibility Theorem in social choice theory: it states that the preferences of individuals in a society cannot in general be converted into a global or 'social' ordering which respects certain minimal requirements for fairness.[17] The fairness requirements give a clear meaning to the optimality of the social ordering, and the theorem states that it is unachievable.

Such examples are also known in economics under the name 'positional goods'. A good such as fame or status consists in a relative position so that the quest for it forms a zero-sum game.[18] For example, it may be possible for anyone to become prime minister, but it is not possible for everyone to become prime minister. 'The best' necessarily – with mathematical necessity – involves compromises.

The first clear use of the contrast between global and local had nothing to do with mathematics, though it was an idea of one of the great mathematicians, Leibniz. Leibniz was a man with a vast range of intellectual interests. One of them was a Christian theory, especially the problem of evil. As most religious people recognize, one of the main difficulties with believing in God is the great amount of evil in the world: if God is good and also all-powerful, how can he allow such horrors? Why does he not make a much better world, if he can? Leibniz's solution to this difficult puzzle, put forward in his book *Theodicy* (Divine Justice) of 1710, is a startling one. God does not create a better world, he says, because there *is* no better world. This is already the best of all possible worlds. The reasons, he thinks, are the same as those just given for the impossibility of a perfect system of tuning or the impossibility of everyone becoming prime minister. That is, it is the (mathematical) impossibility of accomplishing globally what is possible locally. His solution will be examined and defended in the next chapter.

Notes

1 Some preliminary considerations in Lloyd N. Trefethen, *An Applied Mathematician's Apology* (Philadelphia: SIAM, 2022), ch. 6.

2 One likely reconstruction in James Franklin, 'Discrete and Continuous: A Fundamental Dichotomy in Mathematics', *Journal of Humanistic Mathematics* 7 (2017): 355–78, section 2, taken from David H. Fowler, *The Mathematics of Plato's Academy: A New Reconstruction* (Oxford: Clarendon, 1987), 58 and 101.

3 Franklin, 'Discrete and Continuous', section 4.

4 John Dalton, 'Experimental Enquiry into the Proportion of the Several Gases or Elastic Fluids, Constituting the Atmosphere', *Memoirs of the Literary and*

Philosophical Society of Manchester 6 (1805)2: 44–258; Arnold Thackray, 'Documents Relating to the Origins of Dalton's Chemical Atomic Theory', *Manchester Memoirs* 108 (2) (1966): 21–42.

5 Erwin Schrödinger, *Science and Humanism: Physics in Our Time* (Cambridge: Cambridge University Press, 1951), 40.

6 For example, David Crouse and Joseph Skufca, 'On the Nature of Discrete Space-Time, Part 1: The Distance Formula, Relativistic Time Dilation and Length Contradiction in Discrete Space-Time', *Logique et Analyse* 246 (2019): 177–223.

7 Some history in Renaud Chorlay, '"Local-global": The First Twenty Years', *Archive for History of Exact Sciences* 65 (2011): 1–66.

8 Figure 7.1 M.C. Escher's 'Waterfall' © 2024 The M.C. Escher Company The Netherlands. All rights reserved. See www.mcescher.com.

9 Discussion with further examples in James Franklin, 'Mathematical Necessity and Reality', *Australasian Journal of Philosophy* 67 (1989): 286–94.

10 Taken from http://www.optical-illusion-pictures.com/paradox.html.

11 Murray Eisenberg and Robert Guy, 'A Proof of the Hairy Ball Theorem', *American Mathematical Monthly* 86 no. 7 (1979): 571–74.

12 C. B. Allendoerfer, 'Global Theorems in Riemannian Geometry', *Bulletin of the American Mathematical Society* 54 (1948): 249–59.

13 Rudolf Fritsch and Gerda Fritsch, *The Four-Color Theorem: History, Topological Foundations and Idea of Proof* (New York: Springer, 1998).

14 Thomas Reid, *Essays on the Intellectual Powers of Man*, in *The Works of Thomas Reid* (Edinburgh: Maclachlan and Stewart, 1872), vol I, 379.

15 Neil Bibby, 'Tuning and Temperament: Closing the Spiral', in *Music and Mathematics: From Pythagoras to Fractals*, ed. J. Fauvel, R. Flood and R. J. Wilson (Oxford: Oxford University Press, 2003), 13–28.

16 Adam Smith, *The Wealth of Nations* (1776), bk. IV ch. 2.

17 Michael Morreau, 'Arrow's Theorem', *Stanford Encyclopedia of Philosophy* (2014/2019).

18 Fred Hirsch, *The Social Limits to Growth* (London: Routledge and Kegan Paul, 1977); Michael Schneider, 'The Nature, History and Significance of the Concept of Positional Goods', *History of Economics Review* 45 (2007): 60–81.

8

Leibniz's solution to the problem of evil via absolute constraints on God

An excursus on the philosophy of religion may seem out of place in a book about mathematical and similar necessities. But the idea that absolute necessities are so strong as to be not subject to God's will is more than a useful metaphor. Philosophers of faith will better understand what God can and cannot do by appreciating absolute necessities. Philosophers of an atheist persuasion can regard solving the problem of evil as merely an intellectual exercise, but it might not be wise to do so if they have taken the unsolvability of the problem to be a given and the prime reason for dismissing the claims of religion.

God should create the best

There is every reason, initially, to think that if a good and omnipotent God exists, then this must be the best of all possible worlds. It is natural to argue, 'If God is good, then he must be doing his best, and if he is doing his best then this must be the best he can do, and hence the best that is possible.' Even if there were some moral excuses as to why God should be permitted to do less than his best, there is surely something ridiculous or demeaning in the concept of him, as master-craftsman, bothering to create a second-rate world when he could produce a premium-quality article instead. And the more one emphasizes God's transcendence, the less likely it seems that he would indulge in second-rate creations.

Leibniz's theory is that God cannot create a better world because there *is* no better one – that is, that (contrary to appearances) if one part of the world were improved, the changes would necessarily ramify so that the world would be worse elsewhere, and worse overall. It is a 'bump in the carpet' theory: if evil is pushed down here, then it pops up over there. Leibniz put it by saying this is the 'Best of All Possible Worlds'. That phrase proved disastrous for the reputation of his theory, as it suggests the 'Panglossian' (called after Voltaire's caricature) notion that everything is ideal as it is. He does not mean that, but only that designing worlds is much harder than it looks, and that determining the amount of evil in the best one is no easy matter.[1]

In light of modern developments in the mathematics of complex systems, in particular the local/global distinction discussed in the previous chapter, Leibniz's theory is again a credible solution to the Problem of Evil.

Certain clarifications are needed on what 'best' means, such as what God's obligations would be if there were an infinite sequence of increasingly better possible worlds (but no absolute best).[2] The main issue is what 'best' means in the situation where creation is shared between God and free beings. If a best world includes free beings other than God (as it does, on the Leibnizian hypothesis that the best world is this one), then the absolutely best world is the one where all the free beings always choose the best. That may be possible yet beyond God's power to bring about – God's effort to create the best is genuinely hostage to free creatures' efforts to wreck some of his work. As Leibniz puts it, 'God has ordered all things beforehand once for all, having foreseen prayers, good and bad actions, and all the rest.'[3] That is, the world is best not absolutely but given the constraints of human creative acts (some of them bad).

The task of creation has then to be thought of not as resembling the fashioning of a work of art but more as a game of strategy. Creating the universe would be something like playing lightning chess against many opponents, with complete control over one's own pieces but under the necessity of reacting to the other players' choices. Or, like navigation by Google maps, recalculating when the driver goes off-route. The point of these analogies is to instil a sense that even for God, co-creation is a very different task from the creation of a universe under full control. They also give an initial sense of the complexity of the task. Whether God does or does not have knowledge of the future choices of free agents, or of what non-existing free agents would have done in all circumstances if they had existed, the task of working around a series of free decisions made independently by a large number of agents is inherently extremely complex. It is the kind of complexity, as we will see, that is not tamed simply by being omniscient and so able to calculate arbitrarily well.

It is true that God must set a minimum standard as to what worlds should be creatable, and 'pull the plug' if the best possible world falls below that standard. In the last section we will consider the Dostoevskian thought that it is obvious that the empty world is better than this one. Certainly, if that were true, Leibnizian theodicy, his theory of divine justice in creating the world, would be refuted.

Obviously a better world?

The Leibnizian theory's main problem, obviously, is that it is initially very implausible that the world could not be better than it is. We can all think of improvements we would like to make, and that we would make, if it were left up to us. One inevitably thinks along these lines:

> Suppose that we isolate some minor misfortune in the actual world — the stubbing of someone's toe, say — and then we imagine a world that is identical to the actual world, except with that minor misfortune erased. Surely, we would thereby be imagining a better possible world than this one . . .[4]

That argument rests on the assumption that imagination is a reliable guide to possibility, something that has been justifiably doubted in general.[5] But whatever the outcome of discussions on that, the present argument does not depend on any general scepticism about imaginability as a guide to possibility. Nor does it rely on any general argument that 'for all we know, it could be that there are reasons for evils' as in the project of 'sceptical theism'.[6] The present project does agree with sceptical theism in not committing to any particular explanation of why God would permit evil, as does for example the free will defence. But unlike sceptical theism it rests on some particular objections, taken from mathematics, against reliance on imagination as a guide to what is possible specifically in cases of large-scale complex design. The objections do not stem from our mere ignorance, but from our positive knowledge of the ways necessities interconnect in complex structures.

The global/local distinction and constraints on large-scale complex design

The distinction between local and global – between what is possible on a small scale but impossible when everything is fitted together on a large scale – is one of the great themes of mathematics, as described in the previous chapter. The example treated there of tuning and temperament, because of its optimization aspect, is arguably a closer model for Leibniz's theodicy than the Königsberg and four-colour examples. While the latter do illustrate the interaction between local and global possibilities, there is no sense in which some paths across bridges or some map colourings are better than others. But equal temperament includes a definite notion of a best outcome, even though the definition of the best is imprecise (it involves compromises between what is ideal for different instruments and also probably some weighting for simplicity of the outcome). The usual solution of equal temperament is simple and also allows the most important harmonies to be realized with a degree of approximation that does not offend the ear.

The example of a bump in the carpet is even simpler. If the carpet's area is slightly bigger than the room's area, it can be pushed down to make it flat anywhere, but no matter what is done, it cannot be pushed flat everywhere at once. A carpetlayer (or God) could of course make carpet with different properties, but if a carpet has the usual area-constant property, God cannot push it flat any more than a human can, because the necessity is mathematical.

Those are only a few examples. There are many such, well-appreciated by mathematicians, engineers, planners, and architects. Philosophers are unused to them, as most philosophers have never built, let alone designed, so much as a model airplane. They therefore find it easy to assert confidently that 'an all-powerful God, by definition, would be able to achieve the end (the greater good) without using the means (the evil)'.[7] That is too swift. Goods and evils are more intimately and necessarily connected than that. That, indeed, is the real lesson of the 'free will defence', which attempts to explain evils as effects of free will misused, free will itself being a large positive. While it is not remotely plausible that all instances of evil can be so explained,[8] the point is the necessary interconnections of *some* goods and evils. Despite efforts to imagine freely willed choices always turning out well, it is not easy to see how the good of free moral responsibility can be disconnected from the evil of immoral choices. '[God] could have forestalled the occurrence of moral evil only by removing the possibility of moral good.'[9]

The examples, from the strictly mathematical to the free will defence, show that what one thinks naively one can do with complex systems in one's imagination does not translate into what can happen in the world. The essential problem is that imagination itself is local, as the mind can focus on only a small part of a system at once – it can seem to change any part without changing other parts – while design, including the design of worlds and their whole history, is global.

That explains what is wrong with the thought that it is easy to improve the world by removing toe-stubbings while leaving everything else unchanged. As soon as we start making local changes, they ramify and we lose track of what is possible. Design is global, not about tinkering here and there. God's omniscience implies that he has to take into account physics and world history all at once to calculate the best overall world and the total story of that world, satisfying his design criteria.

The scenario of divine choice 'upfront' at the time of creation has become better appreciated through the extensive work in recent decades on the 'fine-tuning' of the constants of nature. According to physics, several constants such as the gravitational constant need to be correct to something like one part in 10^{40} for a physical universe capable of supporting life to be possible. Many other physical facts need to be 'just so' after that to actually achieve life, such as the initial conditions in the Big Bang and the abundance and chemistry of carbon.[10] While there may, or may not, be some distant part of parameter space that also permits life, we know that very little local movement in the actual parameters of the universe is possible without destroying the conditions of life. To the extent that these considerations support the existence of God via an inference to the best explanation, they point to a Leibnizian kind of God, able to calculate to very many decimal places and model the time development of multiple scenarios – God must know the effects of the various tunings of constants in order to be able to choose one in the range that permits life. Setting the laws and initial conditions to achieve not only human life but the best possible history of it, seen as an upfront design problem, resembles adjusting the flow over Niagara to achieve a given distribution of water droplets at the bottom.

What God's design criteria are, we of course do not know well, and we risk making fools of ourselves if we begin speculating. Surely, we can nevertheless make some better suggestions than the sort implicit in claims that he can easily make a better world by removing just our toe stubbing. Perhaps it is like designing a theme park. Or perhaps closer, it is like writing a crime novel: the narrative moves in a realistic context and the author must salt it with apparently trivial incidents which interact with fully drawn characters to eventually come together so as to reveal the criminal and effect the restoration of order. Or . . . but let us not speculate further. If we add in some other design criteria that we might reasonably speculate would be in the mind of God, such as that no-one should be tested beyond their limit, we begin to understand what a computational burden it would be to design a world under conditions of omniscience and how complex we might expect the outcome to be. That is sufficient to indicate why moving to a global view of design takes the focus off the supposed simple advantages of local tinkering.

Plausibility in the light of what? The context of the best-world theory

The Leibnizian solution is certainly not plausible, in isolation. If we just looked around the world, free of preconceptions, and guessed the sort of God that would have designed it (if it were designed at all), we might not take a positive view of divinity. We would be more likely to agree with Hume's suggestion that it is 'the first rude essay of some infant deity'.[11]

But the best-world theory is not intended to be free-standing. The probability of a theory (in the sense of logical probability as in chapter 5) must always be evaluated in the light of one's total evidence. In the case of a philosophical theory such as a theodicy, the total evidence includes such matters as pre-existing reasons for the conclusion (in this case, the conclusion that there is a good creator God); inherent advantages of the theory; the ability to reply to objections; and the credibility of alternative theories (of evil). As with any matter of context, a full treatment would expand indefinitely, but a brief sketch of each will indicate the overall nature of the logical context into which best-world theodicy fits. That will make it possible to understand why it is worthwhile to start to look for excuses for evil in the way a Leibnizian theory does.

Context: Prior reasons for theism

Firstly, the best-of-all-possible-worlds theory only makes sense in the context of prior substantial reasons for believing in a good God. Perhaps one has found the traditional arguments for the existence of God to some degree satisfying. Or some particular religion has seemed a sound candidate for an argument to the best explanation, being a coherent explanation of all facts (including moral ones) superior to alternatives. Given that context, it makes perfect sense to ask if there might be excuses for the evil that God has left in his creation, with a certain bias towards expecting a positive answer. Examining the strength of those arguments is beyond the scope of a theodicy, but their existence must be kept in mind because without that background the project of theodicy is pointless.

Context: Advantages of Best-World Theodicy

Best-world theodicy has a number of inherent advantages. First, best-world theodicy really would *solve* the problem of evil. If God is doing absolutely the best that can be done, he needs no further excuses. Although a few philosophers and theologians have entertained the notion that divine omnipotence includes the power to do the logically and mathematically impossible, it is by and large agreed that locutions such as '2 + 2 = 5' do not name possibilities on which God might choose to confer being. The inability to do the mathematically impossible is not a genuine constraint on omnipotence.

Second, the theory – or actually, even the barest possibility that it is right – is enough to dispose of the 'logical problem of evil',[12] the once well-received idea that there is a strict logical inconsistency in the three propositions:

God is all-good
God is all-powerful
Evil exists.

To show that propositions are logically compatible, all that is needed is to exhibit a *merely possible* scenario in which they are all true. In Leibniz's Best of All Possible Worlds scenario, they *are* all true. Therefore, that deductive form of the problem of evil is refuted.

Third, best-world theodicy makes sense of moral action, including prayer. It might seem at first that Leibniz's theory implies a static world in which neither action nor prayer can improve the world because it is already as perfect as it could be. That is not correct, because world design works in 4D, not 3D: introducing free decisions such as action or prayer into the world changes what is in it, a change which God can take advantage of for his benevolent purposes (either at the time or in the design phase, when he foresees it). God's design works around foreseen decisions and thus incorporates their effects. (Or if God cannot foresee free choices, his creation works by recalculating, like Google maps when the driver veers off-route.)

Fourth – if one is concerned about such things – best world theodicy is consolatory. If someone or their companions are suffering grave evil, it is some consolation that, bad as it is, it is somehow necessary to the design of the universe – a necessary price to pay for the totality of what is good in the universe – and hence not gratuitous. (In the Christian tradition, that applies equally to an evil like the crucifixion. When Jesus says in the Garden of Gethsemane, 'Father, if it be possible, let this cup pass from me' (Matthew 26:39), it may be inferred that, according to Christian theology, as it did not pass, it was not possible.) There is no gratuitous evil in Leibnizian theodicy. It is similar to the situation of a soldier treated as an 'expendable' in battle. If that is due to his general's incompetent or self-aggrandizing strategy, his life is wasted, but if the general planned as well as possible and the soldier was without malice assigned to a dangerous sector of the front just because someone had to be, the soldier can take some comfort that his death is necessary and not wasted. Conversely, if any evil is gratuitous, no such consolation is available.

Context: Least morally offensive theodicy

According to 'antitheodicy', the whole project of theodicy is unfeeling and morally offensive, in that it involves trying to excuse God's allowing the unthinkable such as the Holocaust and children's cancers.[13] Some forms of theodicy do have such a problem, such as 'obstacle-course' theodicies that regard evils as opportunities for the exercise of virtues. But Leibniz's theory does not. It is morally blameless because it excuses God by the absolute impossibility of his choosing any world better than the present one. Necessity is an excuse for actions in general, and absolute necessity is a perfect excuse. No-one can be criticised for failing to exceed the best possible.[14]

Although Leibniz's theory has had a bad reputation on this score, following the success of Voltaire's propaganda, that reputation rests on a caricature of what Leibniz said. According to best-world theodicy, everything is not claimed to be rosy in the best of possible worlds, as Voltaire insinuates. Quite the opposite. God's choice among possible worlds (all possible worlds, including the empty one) is of a kind now familiar in discussion of the Trolley Problem.

In the Trolley case, a driver of a runaway trolley can either allow the trolley to kill five people on one track or divert it onto another track where one person will be killed.[15] Only one uncontroversial aspect of the Trolley problem is relevant here: the forced choice between scenarios that the trolley driver faces plays some exculpatory role.

It is possible to criticize one action of the driver or the other. But we do not say to him – and it would be offensive to say it – 'Whatever you do, you are wrong because you kill someone in order to save others.' Nor do we say that the 'driverdicy' project of excusing the driver is unfeeling and morally offensive because it tries to excuse the inexcusable, namely deliberately allowing one person to die for the sake of a greater good. In a case of forced choice, one of two scenarios must be chosen upfront.

Since God is in the same moral situation (according to the Leibnizian theory), that theory avoids complaints of moral offensiveness due to being utilitarian, in sacrificing some people for the 'greater good' of others. While that is an understandable reaction by those who draw the short straw, it betrays a lack of understanding of what choice between difficult scenarios is like, from the perspective of the moral agent forced to make them. If a general faced with a choice of strategies all involving casualties chooses the one with the least number of casualties, the natural grudges of those soldiers who lose out can coexist with an acceptance that the general could not have done morally better, since any other decision would have had the same problem (but worse). To describe the best choice of a spectrum of choices, all involving evils, as 'doing evil that good may come of it' would fail to appreciate the nature of upfront choice at the design phase.

Context: Do alternatives to personal theism trivialize evil?

A last matter of context is whether alternative cosmological theories to personal theism have an adequate sense of evil themselves. Some have thought that materialist atheism, the leading alternative to the personal theism which Leibniz's theory defends, trivializes evil by regarding persons as mere collocations of matter.[16] If it did, that would undermine the robust sense of the seriousness of terrible evil which gives point to the problem of evil. That would be especially so if the atheist were so determinedly naturalist as to accept Mackie's 'error theory' of ethics, according to which any ethical concept is too 'queer' to fit into a scientific picture of the universe.[17] Atheists advancing the problem of evil are naturally keen to emphasize how terrible evils are, but if there is any suspicion that their final position might undermine the objective badness of evil, questions could be asked. What about alternatives to materialism and personal theism such as pantheisms and emergentisms?[18] Are they in a different position when it comes to the seriousness of evil suffered by humans? Those questions are difficult and wide-ranging and no attempt to address them will be made here. But it needs to be kept in mind that if evil is a problem for personal theism, it may also be one for alternatives. That is a context to be taken into account when evaluating Leibnizian theodicy.

These matters of context create a framework into which best-world theodicy fits naturally, if only it can be made to work by explaining why evils and goods can be expected to be inextricably linked. As argued above, that can be done.

Dostoevskian choice and the empty world

Finally, the Leibnizian theory faces the objection that there is a world that is obviously possible and obviously better than this one, namely the empty one (that is, empty of everything except God and any other necessary beings that there may be). A strength of Dostoevsky's presentation of that objection is that he does see it in terms of upfront choice between scenarios: 'Imagine that you are creating a fabric of human destiny with the object of making men happy in the end . . . but that it was essential and inevitable to torture to death only one tiny creature . . . would you consent to be the architect on those conditions?' . . . "No, I wouldn't consent."'[19]

That is a powerful argument. But there are two considerations against it. The first is that it is far from obvious that the empty world is possible. With issues like the possibility of whole universes, we are so far from our cognitive home ground that we may well doubt if our modal intuitions founded on ordinary experience have any purchase, and whether our apparent ability to imagine an empty world is here a reliable guide to possibility. The Neoplatonists denied that was possible, holding that God necessarily 'overflowed' into creation,[20] and it is hard to see how we could know that is false.

Secondly, on reflection, we may not be so sure that we would make the choice Dostoevsky suggests, if we were put to the test and were fully informed. Suppose that we were given the choice of pushing a button so that, not only were we never born, but the whole world was never born. How sure would we be about pressing it? It would be very tempting, as we considered the evils we would prevent. But turning over in our minds what would be lost might be enough to cause hesitation. We would find ourselves comparing scenarios, as Leibniz says God does in designing the world.

We should remember, too, that with respect to part of the future we are ourselves granted Dostoevsky's choice – namely, that part consisting of our own descendants. If we assume, as is inductively reasonable, that future people will have much the same balance of goods and evils as present ones, it is a real question whether we should have children. David Benatar argues in *Better Never to Have Been* that we should make the Dostoevskian choice and confer on our potential descendants the benefit of non-existence.[21] The reaction to his proposal has been muted rather than enthusiastic. Even in philosophical circles, the birth of a healthy baby is not normally considered a cause for mourning. So, while it is almost philosophical orthodoxy to congratulate oneself on one's sympathy for human suffering in agreeing with Dostoevsky, when philosophers are presented with the opportunity to partially put into effect the Dostoevskian choice, almost all stay their hands.

It is not the final judgement that is significant. Benatar's position is arguable and not obviously wrong. What is significant is the hesitation over it. To hesitate in the face of that decision is to start weighing scenarios morally, just as Leibniz suggests God does at creation. Once we begin doing that, we must consider the known ways in which goods and evils are necessarily interconnected, as discussed above.

Conclusion

It is possible to defend the natural thought that God ought to be doing his best and so this world is the best he can do. As Leibniz explains, that is not done via any general scepticism about our knowledge of God and creation but requires an appreciation of the mathematical absolute necessities involved in any instance of complex design. They are especially difficult in the largest-scale design project possible, a whole world that needs to be created as a platform for a large population of interacting free decision-makers. Given the context of the problem, Leibniz's view that this is the least worst of all possible worlds remains a live option.

Notes

1 James Franklin, 'Two Caricatures II: Leibniz's Best World', *International Journal for Philosophy of Religion* 52 (2002): 45–56.
2 Further in James Franklin, 'The Global-local Distinction Vindicates Leibniz's Theodicy', *Theology and Science* 20 (4) (2022).
3 Gottfried Leibniz, *Theodicy: Essays on the Goodness of God, the Freedom of Man and the Origin of Evil* (1710), trans. E. M. Huggard, ed. Austin M. Farrer (London: Routledge and Kegan Paul, 1951), par. 9, p. 128.
4 Campbell Brown and Yujin Nagasawa, 'The Best of All Possible Worlds', *Synthese* 143, no. 3 (2005): 309–20, at 310.
5 E.g. Stephen Yablo, 'Is Conceivability a Guide to Possibility?', *Philosophy and Phenomenological Research* 53 (1993): 1–42.
6 Michael J. Almeida and Graham Oppy, 'Sceptical Theism and Evidential Arguments from Evil', *Australasian Journal of Philosophy* 81 (2003): 496–516, review in Timothy Perrine, 'Skeptical Theism', *Stanford Encyclopedia of Philosophy* (2023), https://plato.stanford.edu/entries/skeptical-theism/ (which do not mention Leibniz or global/local considerations).
7 Mark Piper, 'The Perennial Problem of Evil', *Think* 2, no. 4, (2003): 65–73, at 70.
8 D. Z. Phillips, *The Problem of Evil and the Problem of God* (London: SCM Press, 2004), ch. 4.

9 Alvin Plantinga, *God, Freedom and Evil*, 2nd edn (Grand Rapids, MN: Eerdmans, 1977), 30.

10 Martin Rees, *Just Six Numbers: The Deep Forces That Shape the Universe* (London: Weidenfeld and Nicholson, 1999); Geraint F. Lewis and Luke A. Barnes, *A Fortunate Universe: Life in a Finely Tuned Cosmos* (Cambridge: Cambridge University Press, 2016).

11 David Hume, *Dialogues Concerning Natural Religion* (1779), part 5.

12 James R. Beebe, 'Logical Problem of Evil', *Internet Encyclopedia of Philosophy*, (2003, updated 2005), http://www.iep.utm.edu/evil-log/

13 E.g. N. N. Trakakis, 'Antitheodicy', in *Blackwell Companion to the Problem of Evil*, ed. J. P. McBrayer and D. Howard-Snyder (Oxford: Blackwell, 2013), 363–76.

14 James Franklin, 'Antitheodicy and the Grading of Theodicies by Moral Offensiveness', *Sophia* 59 (2020): 563–576.

15 Judith Jarvis Thomson, 'Killing, Letting Die and the Trolley Problem', *Monist* 59 (1976): 204–17.

16 Discussion in Erik Wielenberg, *Value and Virtue in a Godless Universe* (Cambridge: Cambridge University Press, 2005).

17 Nicholas L. Sturgeon, 'Evil and Explanation', *Canadian Journal of Philosophy*, 25 (Supplementary Volume 1), (1995): 155–85.

18 E.g. James Franklin, 'Emergentism as an Option in the Philosophy of Religion: Between Materialist Atheism and Pantheism', *Suri: Journal of the Philosophical Association of the* Philippines 7, no. 2 (2019): 1–22.

19 Fyodor Dostoevsky, *The Brothers Karamazov* (1880), bk 5, ch. 4.

20 Pseudo-Dionysius (c. 500), *On the Divine Names*, in Dionysius the Areopagite: *On the Divine Names and the Mystical Theology*, trans. C. E. Rolt (London: SPCK, 1920), 4.10, pp. 101–2 and 13.1, pp. 184–5.

21 David Benatar, *Better Never to Have Been: The Harm of Coming into Existence* (Oxford: Oxford University Press, 2006).

Part Five

Ethical Necessities Are Absolute

9

Ethics: The parallel with mathematics

As described in the last few chapters, mathematics and logic are the home ground of absolute necessities, as has always been recognized. Ethics is much less so, even though many have sought an objective grounding for it in one way or another. Before developing an objectivist view of ethics based on absolute necessities in the next chapter, we will examine the parallel between ethics and mathematics. The two are much closer to each other than either is to the natural or social sciences.

Objectivity in ethics and mathematics

For those seeking to defend the objectivity of ethics within a realist framework, emphasizing its parallels with mathematics is an attractive line of argument. Arguments for ethical relativism arising from the mere fact that ethical principles are held by people and have evolved through natural causes face the objection that mathematical truths do not have their objectivity impugned by similar considerations.[1] The fact that ethical knowledge (if there is any) seems to be *a priori*, arising from pure thinking rather than measurement or scientific observation, is also something it shares with mathematical knowledge,[2] so cannot, in itself, tell against the objectivity of ethics.

Some obstacles certainly stand in the way of using the parallel with mathematics to bolster objectivist views of ethics. Some philosophies of both ethics and mathematics undermine the parallel. For example, non-cognitivist philosophies of ethics that see it as fundamentally a set of techniques about action developed for evolutionary and social reasons, as 'traffic rules for self-assertors',[3] deny that ethics is a body of truths about a subject matter at all. From the other direction, some popular nominalist philosophies of mathematics also deny that mathematics is a genuinely contentful science of a subject matter.

At one time Peter Singer argued:

> The defenders of ethical intuitionism argued that there was a parallel in the way we know or could immediately grasp the basic truths of mathematics: that one plus one equals two, for instance. This argument suffered a blow when it was shown that the self evidence of the basic truths of mathematics could be explained in a different and more parsimonious way, by seeing mathematics as a system of tautologies, the basic elements of which are true by virtue of the meanings of the terms used. On this view, now widely, if not universally, accepted, no special intuition is required to establish that one plus one equals two – this is a logical truth, true by virtue of the meanings given to the integers 'one' and 'two', as well as 'plus' and 'equals'. So the idea that intuition provides some substantive kind of knowledge of right and wrong lost its only analogue.[4]

However, that kind of philosophy of mathematics is not universally or even widely accepted. In fact, it has had hardly any adherents in philosophy since Frege and Russell found it untenable over a hundred years ago.[5] The persistence of such views, however, does show that realist philosophies of mathematics cannot be taken for granted any more than realist philosophies of ethics can be. Indeed, one could in principle use the mathematics-ethics analogy to support an anti-realist or pluralist view of both.[6]

Mathematics and Eternal Verities: Realist Philosophies of Mathematics

Traditionally, mathematics has been taken to be the study of eternal truths about an objective subject-matter. The incommensurability of the diagonal and side of a square, the logarithmic decay of the density of primes, and the existence of exactly 26 sporadic finite simple groups, appear to be unchanging, provably true facts about a terrain of mathematical entities that exists independently of human actions, knowledge or wishes. As Plato says, it would be ludicrous to interpret mathematicians' talk of actions like adding numbers, constructing circles and extracting square roots as implying that mathematics is really about action. It is pure knowledge, 'the knowledge of that which always is, and not of a something which at some time comes into being and passes away'.[7] Arithmetic is normative for the actions of adding up by accountants, but it is itself neither essentially normative nor about actions.

As we saw above, realist philosophies of mathematics offer various accounts of what eternal verities mathematics is, in fact, about (in contrast to nominalist philosophies which represent mathematics as a mere language of science, or method of deriving one substantive scientific truth from others, or manipulation of uninterpreted symbols). The most prominent realist philosophy of mathematics is Platonism, according to which numbers, sets and the like are classical 'abstract objects': acausal, atemporal objects existing necessarily in a Platonic realm, neither physical nor mental. Since Platonism (in and outside the philosophy of mathematics) is subject to several classic objections, the alternative realist philosophy of mathematics, Aristotelianism, is preferable. It takes mathematics to be as necessarily true and contentful as does Platonism but also takes it to be about properties of the physical world (and any other non-abstract world there may be) – properties such as symmetry, ratio, pattern, continuity, and flow. That philosophy was developed in Chapter 3.

What one learns from mathematical examples like the Königsberg bridges is that there exists a world of necessities, understandable by pure thought, which have consequences in constraining the real (non-abstract) world. Mathematical necessities do not live just in a Platonic realm of 'pure' mathematics but in our physical world. That opens one's mind to becoming attuned to the necessities of ethical reality.

That is not to claim that ethical necessity and mathematical necessity are identical. If it is mathematically impossible to walk over all the bridges just once, one does not have the choice of doing it. But if it were impossible to walk over all the bridges justly, one could still choose to walk over them. Ethical necessities constrain what it is possible to do morally, which is not the same as constraining what it is possible to do mathematically or physically. But then one can also ignore mathematical necessities if one chooses. For example, the fact that 1 + 1 = 2 does not constrain accountants to always add $1 and $1 and get $2. Sound accounting conforms to that addition, but 'creative' accounting does not.

Ethics and eternal verities: The good versus the right

Traditionally, ethics has not been understood as the study of eternal truths about an independently existing objective subject matter in the way mathematics has been. It has been usual to see ethics as (solely or largely) about human actions – their rightness, their obligatoriness, what is a duty and what is permitted, which are virtuous. Even realist ethical positions that take it to be an objective matter which actions are right, such as Kantian universalizability or classical utilitarianism, suppose that it is fundamentally actions (or action-orientated virtues) that exhibit moral qualities such as rightness.

But to appreciate what Plato's perspective has to offer in comparing mathematics with ethics, we need to follow him in seeing what is ethically admirable or criticizable as in the first instance something prior to actions. His choice is that ethics is centrally about excellence of character (*arete*) and the 'just' ordering of the state. We need not accept those particular choices, much less Plato's views on what constitutes excellence, to understand the general point that ethics is not *fundamentally* about what to do – even though the fundamental ethical facts have consequences for what to do and right action is certainly an important part of ethics.[8]

There are two basic reasons for thinking that. Firstly, what we are most disturbed by ethically – what most violently forces itself on us as ethically objective – is not anything to do with actions, but the terribleness of suffering. What makes something a tragedy is first and foremost the happening of serious evil to some being of worth.

Secondly, whenever we ask *why* some action is right or wrong, we find we are led back to reasons that are not themselves about action but concern the good or evil of those affected. For example, what makes the act of killing wrong is in the first place the evil of the death of the victim (rather than the action's violating some rule or being contrary to some virtue). That explains why the rule against killing can be relaxed in the case of killing in self-defence, since then there is a conflict between the evils of the death of the victim and the death of the attacker.[9]

While the emphasis on actions and their evaluation has dominated ethics, two of its leading figures, Kant and Moore, stood for a position where the 'good' (what was of moral worth in itself) contrasted with the 'right', which concerned actions. Moore states the contrast as a main theme of his *Principia Ethica*:

> I have tried in this book to distinguish clearly two kinds of question, which moral philosophers have always professed to answer, but which, as I have tried to shew, they have almost always confused both with one another and with other questions. These two questions may be expressed, the first in the form: What kind of things ought to exist for their own sakes? the second in the form: What kind of actions ought we to perform?[10]

Kant too, though discussing at length duties, rules and respect, sees at the bottom of ethics something not itself about actions, namely the dignity or worth of persons, which 'exacts' respect and duties:

> Man regarded as a *person*, that is, as the subject of a morally practical reason, is exalted above any price, for as a person he is not to be valued merely as a means

> to the ends of others, or even to his own ends, but as an end in himself, that is, he possesses a *dignity* (*Würde*) (an absolute inner worth (*Werth*)) by which he exacts *respect* for himself from all other rational beings in the world. He can measure himself with every other being of this kind and value himself on a footing of equality with them.[11]

If 'what ought to exist for its own sake' and 'absolute inner worth' are objective notions at all, as Moore and Kant think, they must be the kinds of eternal verities of which Plato speaks, ethical counterparts of the verities that mathematics deals in. The same is true, indeed, of the ethical ideal of classical utilitarianism – it says that the action that contributes to the greatest happiness of the greatest number is right *because* the greatest happiness of the greatest number is an external absolute ideal (itself unrelated to action). In general, there are many things which are ends of action because we take them to be valuable in themselves – survival, health, artworks, the preservation of ecosystems, tenure. Aristotelian ethicists particularly emphasize the perfecting of human rational powers as a good for humans[12] – and the notion of a perfection is one that is both paradigmatically ethical and paradigmatically not about action. The inherent value of these goods is what gives actions that promote them their ethical point.

A realist option in the philosophy of the ethical good is developed in the next chapter, based on the inherent worth of persons. It will explain where the necessities of ethics come from. But before that it is useful to deepen the parallel between ethics and mathematics by looking at the role of equality in each.

Equality and symmetry arguments in mathematics

Equality is, paradigmatically, a central concept of mathematics. Mathematical writing is full of equations. Symmetry, central to algebra and mathematical physics, concerns the equality of parts of a system (in some respect such as shape). To have bilateral symmetry (the simplest form of symmetry) is just to consist of two parts equal in some respect; for example, to be a palindrome is to have the second half of the text repeating the first in the opposite order.

The ability of arguments from symmetry to generate contentful propositions of (applied) mathematics is well illustrated in Archimedes' derivation of the law of the lever, an example which much impressed the creators of the Scientific Revolution and led to successful efforts to emulate it and thus mathematize physics.[13] Following it in detail (in a simple case) is useful for Platonic training of the soul in a mathematical necessity.

The law of the balance states that two weights on a beam supported by a fulcrum balance each other if the ratios of the weights is the inverse of the ratios of their distances from the fulcrum. That is, lighter weights far away can balance heavy weights closer in. Archimedes does not regard this as a physical law needing experimental support but offers to demonstrate it, in the manner of Euclid, from absolutely certain first principles. In the simplest example, where one weight is twice the other, his proof works like this:

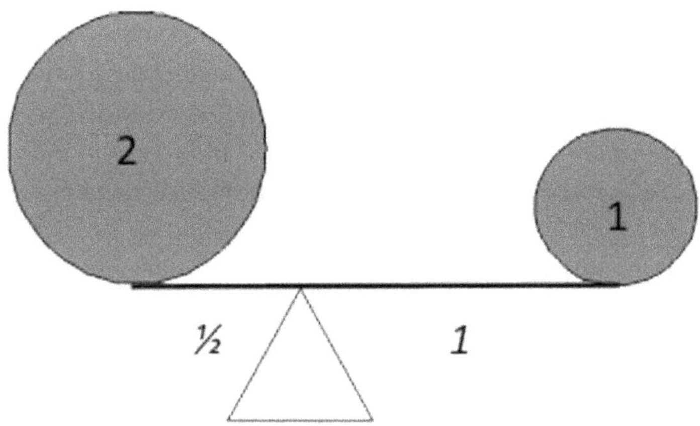

Figure 9.1 A weight of 2 close to the fulcrum balances a weight of 1 twice as far away.

To prove that weights so positioned do balance, Archimedes first extends the (weightless) beam in each direction so that the fulcrum is now in the middle (the weights do not change position).

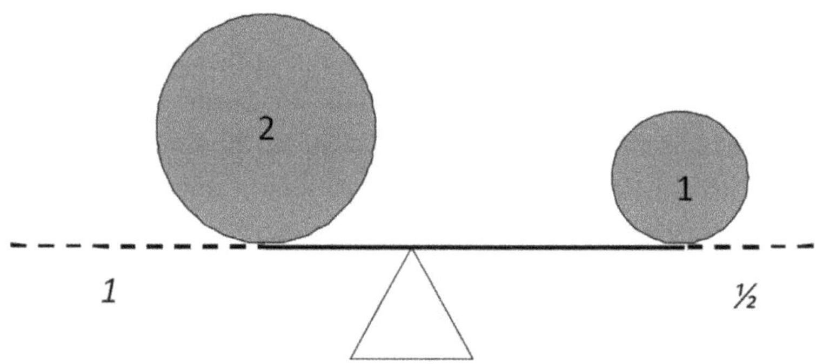

Figure 9.2 The beam is extended so that the fulcrum is in the middle.

The weights are now imagined as malleable, like clay, and are gently patted down to rest with uniform thickness on the beam, as in Figure 9.3.

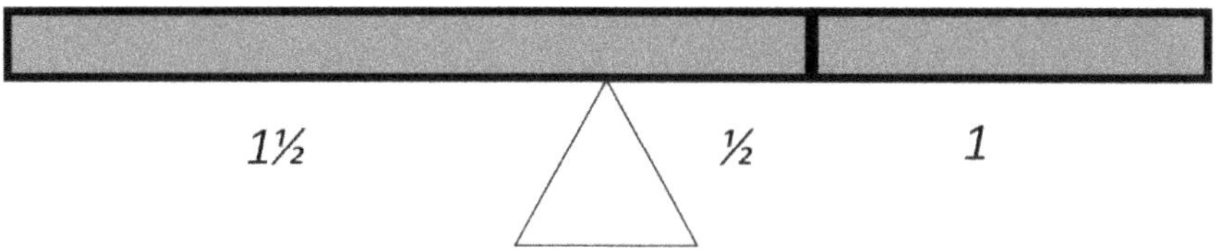

Figure 9.3 The weights are patted down to rest uniformly on the beam.

We see that, in the final state, weight is uniformly distributed, with equal amounts on each side of the fulcrum. So, by symmetry, the weights must balance, therefore the original pair of weights must also balance. All there is at the bottom of the argument is the equality of the pieces on each side of the fulcrum, which causes them to balance; 'equal weights at equal distances are in equilibrium'. That obvious and apparently innocuous principle of equality is parlayed by Archimedes' genius into a complex theory of contentful propositions about the balancing of unequal weights at unequal distances.[14]

The role of symmetry in mathematical physics has only increased since then, to the extent that 'In the latter half of the 20th century symmetry has been the most dominant concept in the exploration and formulation of the fundamental laws of physics.'[15]

A mathematical example with perhaps even closer analogies to ethics is probability (that is, factual probability, not the logical probability that was the main topic of chapter 5). As is well known, complicated questions about the probability of events in games with dice and cards are solved by counting the numbers of equiprobable outcomes. The probability of two dice giving a total of 2 (that is, of both showing 1) is 1/36, while the probability of their giving a total of 3 (that is, one of them showing 1 and the other 2) is 2/36. The reason is that the first event consists of one of the 36 basic equiprobable outcomes of two dice (1 and 1), while the second event consists of two of them (1 and 2, and 2 and 1). It is the equiprobability of the basic outcomes that reduces problems in probability to the pure mathematics of counting. The equiprobability of the basic outcomes, it is agreed, results from a symmetry argument. Outcomes such as the 36 possible falls of two dice are equiprobable because there is in some sense a symmetry between them. Debate has been heated as to what this symmetry consists in – is it the physical symmetry of the dice? The equality of the long-run-observed relative frequency of outcomes? Our equal ignorance of the outcomes? Those are fair questions, but the calculation of outcomes does not depend on answering them. Provided the equiprobability of the basic outcomes is granted, they can be counted to give the probabilities of combinations of them.

As Pascal, who discovered mathematical probability, said, it is surprising that chance is subject to such strict and exact mathematical methods: 'by thus uniting the demonstrations of mathematics to the uncertainty of chance, and reconciling what seem contraries, it can take its name from both sides, and rightly claim the astonishing title: the Geometry of chance (*aleae Geometria*)'.[16] The details of his symmetry argument will be considered more closely in Chapter 11.

The ability of symmetry principles to produce contentful truths about the worlds of physics and probability can act, in Platonic fashion, as an inspiration for ethics. If it is surprising that ethics is subject to symmetry arguments, it is no more so than the fact that chance is.

Equality and symmetry arguments in ethics

Ethics can follow applied mathematics into objectivist territory by also relying on symmetry arguments, or deductions from equality, provided it can give some minimal account of why the equality holds.

The best-known example of what might be called the 'Archimedean spirit' in ethics, of deducing a complex ethical theory from a symmetry principle, is Rawls' *Theory of Justice*. Rawls regards distributive justice as dealing with 'the way in which the major social institutions distribute fundamental rights and duties and determine the division of advantages from social cooperation'.[17] He proposes to deduce just distributive arrangements from some assumptions about an 'initial position', in which individuals must choose principles from behind a 'veil of ignorance', which allows them self-interest and knowledge of general facts about human nature, but no knowledge of what position in society they will be born into. Rawls' model of deduction is a Euclidean one. 'We should strive for a kind of moral geometry', he writes, 'with all the rigor which this name connotes'.[18]

Fundamental to the principles is equality: they are 'the principles that free and rational persons concerned to further their own interests would accept in an initial position of equality as defining the fundamental terms of their association'.[19]

Rawls denies to the participants in the initial position any substantive moral knowledge or conceptions of the good life. That throws even more weight on the pure symmetry between participants as bare moral agents, as Rawls recognizes. He writes,

> the parties are symmetrically situated in the original position. This models our considered conviction that in matters of basic political justice citizens are equal

> in all relevant respects: that is, that they possess to a sufficient degree the requisite powers of moral personality and the other capacities that enable them to be normal and fully cooperating members of society over a complete life. Thus, in accordance with the precept of formal equality that those equal (similar) in all relevant respects are to be treated equally (similarly), citizens' representatives are to be situated symmetrically in the original position.[20]

On the question of where equality comes from, Rawls' official position is that his concept of equality is 'political, not metaphysical'. Thus (as in discussions of the meaning of probability), he aims to retain the obviousness of and widespread agreement on equality without going into difficult discussions on what it is about humans that makes them equal. However, avoiding foundational inquiry entirely is rarely possible in philosophy and Rawls does have a brief discussion of the 'basis of equality' which gives a sketch of his answer. Asking why animals do not count as participants in the original position, he says that 'equal justice is owed to those who have the capacity to take part in and to act in accordance with the public understanding of the initial situation'.[21] His use of 'capacity' to extend rights beyond those who can actually understand to infants and the severely disabled indicates a degree of metaphysical stance on human nature.

Rawls' ambition is restricted to explaining political arrangements. It does not extend to ethics in general, such as the ethics of ordinary interpersonal relationships. Plainly bare and impersonal principles of equality are more suited to the discussion of political arrangements than to loving relationships between family and friends. Nevertheless, if we do look beyond the political realm, the principle of equality interacts with more substantive ethical principles and moderates how they work. We will consider a few brief examples.

Love, in its various senses, is central to ethics and is necessarily particular. Yet Jesus's call to 'love your neighbour as yourself' expresses a widely felt urge to include all within the ambit of loving behaviour, such as charity, as far as possible. Natural and ethical ways in which one prefers and acts on behalf of one's family are properly restricted by laws against nepotism.

The substantive right to an education, based on allowing human rationality to be realized, implies educational actions, but not necessarily identical actions for each child. The circumstances of people matter and they enter into the calculation. An equal right of children to a fair share of educational resources, for example, will require different actions in the cases of a musical or mathematical prodigy, a well-adjusted child of average intelligence, and an intellectually disabled child. All have rights to education, but the plans must be tailored to each child's abilities to profit from teaching, and one plan may cost more than another. That is typical of how variable circumstances interact with equality. (In Aristotle's mathematical language, the just distribution is proportional rather than numerically equal.[22])

Naturally, that leads to long debates on the detailed consequences of the principle of equality with, for example, a need to resolve the tension between equality of outcomes and equality of opportunity, what weight should be given the claims of the worst off, and the like. Those debates cannot be settled simply by appealing the equality of persons. Nevertheless, the ideal of equality is not vacuous. Appeals to equality of consideration are always very powerful. And that does not mean merely that equality is weighted heavily in comparison with other considerations. It means that any other consideration, such as skin colour or age or wealth is, by default, of absolutely no weight, and the moral relevance of any consideration must be established in the face of the strong presumption against its relevance. Thus, in the 1858 Lincoln-Douglas debates on slavery, Stephen Douglas was probably right as a matter of historical fact to claim that the writers of the Declaration of Independence had only had white men in mind in proclaiming 'all men are created equal', but Abraham Lincoln's case that the logic implied equality for all prevailed.[23]

'Our' (post-Christian and post-Enlightenment) ethics has become egalitarian in principle, to the extent that 'The principle of equal dignity and respect is now accepted as a minimum standard throughout mainstream Western culture'.[24] But it should not be forgotten that ethics without a principle of the equality of persons is possible and has been commonplace historically. Ancient Greek ethics was barely aware of such a principle, so the ethics of Plato and Aristotle is alien to ours. Slave-owning societies, of course, recognized no such principle. In contemporary ethics, the most prominent denier of the principle is Peter Singer. Singer's ethics is very mathematical, rigorously deduced from *a* principle of equality, but not the equality of persons. It rests on the 'basic principle of equality: the principle of equal consideration of interests'. All interests are equal; 'an interest is an interest, whoever's interest it may be'.[25] Thus, an animal's pain is entitled to equal consideration to an equal pain in a human. That principle has been responsible for Singer's controversial and radical departures from traditional ethics, such as his defence of the permissibility of infanticide for healthy babies. 'The preference utilitarian reason for respecting the life of a person cannot apply to a newborn baby. A newborn baby cannot see itself as a being which might or might not have a future, and so cannot have a desire to continue living.'[26] Singer's preference utilitarianism counts interests (as equal) in abstraction from the beings whose interests they are, and the latter do not count morally in themselves.

These dramatic implications of divergence from a principle of equality of persons reveal the moral significance of the principle itself. It is neither meaningless nor vacuous nor infinitely flexible. Nor is it widely agreed, if we speak cross-culturally and cross-temporally.

Conclusion

W. D. Ross wrote in 1930:

> That an act, *qua* fulfilling a promise, or *qua* effecting a just distribution of good, or *qua* returning services rendered, or *qua* promoting the virtue or insight of the agent, is *prima facie* right . . . is self-evident just like a mathematical axiom, or the validity of a form of inference, is evident. The moral order expressed in these propositions is just as much part of the fundamental nature of the universe (and, we may add, of any possible universe in which there were moral agents at all) as is the spatial or numerical structure expressed in the axioms of geometry or arithmetic. In our confidence that these propositions are true there is involved the same trust in our reason that is involved in our confidence in mathematics; and we should have no justification for trusting it in the latter sphere and distrusting it in the former. In both cases we are dealing with propositions that cannot be proved, but that just as certainly need no proof.[27]

Since then, many relativist and historicist currents have undermined that kind of robust belief in the absolute objectivity of ethical truths. Those currents have had less success with undermining belief in mathematical truths while earlier philosophies of mathematics that dismissed mathematics as a mere language or heap of tautologies have waned. Leading philosophies of mathematics are realist, supporting Ross's view that the mathematical structure of the universe is part of its fundamental and necessary nature and that the most central truths describing it are knowable with certainty. The parallels between mathematics and ethics suggest that the same can be said of the ethical structure of reality, thus reviving the essentials of Plato's view that mathematics is the ideal body of knowledge for inducing an understanding of ethical truths in the soul.

Notes

1 Jonathan Lear, 'Ethics, Mathematics and Relativism', *Mind* 92 (1982): 38–60; James Franklin, 'On the Parallel between Mathematics and Morals', *Philosophy* 79 (2004): 97–119.

2 Christopher Peacocke, 'Moral Rationalism', *Journal of Philosophy* 101 (2004): 499–526.

3 Annette C. Baier, 'What Do Women Want in a Moral Theory?' *Noûs* 19 (1985): 53–63.

4 Peter Singer, 'Introduction', in *Ethics*, ed. Peter Singer (Oxford: Oxford University Press, 1994), 3–13, but largely withdrawn in Katarzyna de Lazari-Radek and Peter Singer, *The Point of View of the Universe: Sidgwick and Contemporary Ethics* (Oxford: Oxford University Press, 2014), chs. 3–5.

5 Alan Musgrave, 'Logicism Revisited', *British Journal for the Philosophy of Science* 28 (1977): 99–127, section 2; Neil Tennant, 'Logicism and Neologicism', *Stanford Encyclopedia of Philosophy*, (2013/2023), section 6.

6 For example Justin Clarke-Doane, *Morality and Mathematics* (Oxford: Oxford University Press, 2020); Justin Clarke-Doane, 'The Ethics-mathematics Analogy', *Philosophy Compass*, 15, no. 1 (2020), e12640; some contributions in *Explanation in Ethics and Mathematics: Debunking and Dispensability*, ed. U. B. Leibowitz and N. Sinclair (Oxford: Oxford University Press, 2016); reply in James Franklin, '"Let No-one Ignorant of Geometry. . .": Mathematical Parallels for Understanding the Objectivity of Ethics', *Journal of Value Inquiry* 57 (2023): 365–84, section 5.

7 Plato, *Republic* 527a–b.

8 James Franklin, *The Worth of Persons: The Foundation of Ethics* (New York: Encounter Books, 2022), ch. 1.

9 Franklin, *The Worth of Persons*, ch. 2.

10 G. E. Moore, *Principia Ethica* (Cambridge: Cambridge University Press 1903), preface.

11 Immanuel Kant, *The Metaphysics of Morals* (1797), 6:435, trans. M. Gregor (Cambridge: Cambridge University Press, 1991), 186; this realist interpretation of Kant defended in Allen Wood, *Kant's Ethical Thought* (New York: Cambridge University Press, 1999), e.g. 157–8.

12 John Hacker-Wright, 'Virtues as Perfections of Human Powers: On the Metaphysics of Goodness in Aristotelian Naturalism', *Royal Institute of Philosophy Supplements* 87 (2020): 127–49.

13 Epaminondas Vampoulis, 'Archimedes in Seventeenth Century Philosophy', in *The Genius of Archimedes: 23 Centuries of Influence on Mathematics, Science and Engineering*, ed. S. A. Paipetis and M. Ceccarelli, (Dordrecht: Springer, 2010), 331–43.

14 Archimedes, *On the Equilibrium of Planes*, in *The Works of Archimedes*, ed T. L Heath (Cambridge: Cambridge University Press, 1897), 189–220, book 1 proposition 6;

discussion in James Franklin, 'Early Modern Mathematical Principles and Symmetry Arguments', in *The Idea of Principles in Early Modern Thought*, ed. P. Anstey (New York: Routledge, 2017), ch. 1.

15 David J. Gross, 'The Role of Symmetry in Fundamental Physics', *Proceedings of the National Academy of Sciences* 93 (1996): 14256–9.

16 Blaise Pascal, Address *Celeberrimae Matheseos Academiae Parisiensi* (1654), in *Oeuvres Complètes*, ed. J. Mesnard (Bruges: Desclée de Brouwer, 1964–70), vol. 2, 1034–5.

17 John Rawls, *A Theory of Justice* (revised edn, Oxford: Oxford University Press, 1999), 6.

18 Rawls, *Theory of Justice*, 105, though he admits ethics is not in that state yet: discussion in Michael B. Gill, 'Morality is not like Mathematics: The Weakness of the Math-moral Analogy', *Southern Journal of Philosophy* 57 (2019): 194–216.

19 Rawls, *Theory of Justice,* 10.

20 John Rawls, *Justice as Fairness: A Restatement* (Cambridge, MA: Harvard University Press, 2001), 18.

21 Rawls, *Justice as Fairness,* 442; full discussion in Jeremy Waldron, *One Another's Equals: The Basis of Human Equality* (Cambridge, MA: Harvard University Press, 2017).

22 Aristotle, *Nicomachean Ethics* 1130b–1132b.

23 Discussion in Peter Westen, 'The Meaning of Equality in Law, Science, Math, and Morals: A Reply', *Michigan Law Review* 83 (1983): 604–63.

24 Stefan Gosepath, 'Equality', *Stanford Encyclopedia of Philosophy* (2001/2007), section 2.3, with brief history, archived at https://plato.stanford.edu/ARCHIVES/SPR2013/ENTRIES/equality/

25 Peter Singer, *Practical Ethics* (2nd edn, Cambridge: Cambridge University Press, 1993), 21.

26 Singer, *Practical Ethics,* 171.

27 W. D. Ross, *The Right and the Good* (Oxford: Clarendon, 1930), 29–30.

10

The foundation of ethics in the worth of persons

'What a piece of work is man', says Hamlet. 'How noble in reason, how infinite in faculty, In form and moving how express and admirable, In action how like an Angel, In apprehension how like a god . . . .' Hamlet is right that possession of such properties as reason, apprehension, and the capacity to act confer on humans a moral weight, a nobility, dignity or worth that makes it important, absolutely speaking, what happens to humans and what they do. When we see pictures of atrocities, we understand that the victims were people like us, people who, until their end, possessed the same reason, apprehension, personal unity, hope, love, suffering, and individuality as ourselves.

What happens to humans can be a tragedy. What happens to rocks cannot. That is what makes their death a tragedy and not, like the explosion of a lifeless galaxy, a firework. Gross metaphysical differences such as those between humans and rocks imply gross moral differences.

If we turn to philosophical works on the foundations of ethics, the array of concepts is very different from talk of nobility, perfections, dignity, and tragedy. Those are found in classic literary texts, but in ethical works, discussion is all about right and wrong actions, consequences and rules, rights and duties, virtues and dilemmas. Those are all respectable ethical topics, but they are only central to ethics in a view of it as fundamentally about what to do (how to live, what actions are right and wrong). That is a very partial view of ethics. It skates over the surface, by disconnecting what humans ought to do from what sort of entities they are. Its superficiality often leads to painting ethics in pastel shades, as if it is a polite and reasoned discourse about subtle questions, infused with little sense of urgency.

Basing Ethics on the Worth of Persons

Ethics is not fundamentally about what to do. Of course, questions of right and wrong, good and bad actions, rights and duties, and virtues and vices, are an important part of ethics. But they are not fundamentally what ethics is about. That is for two reasons. Firstly, what we are most disturbed by ethically – what most violently forces itself on us as ethically objective – is not anything to do with actions, but the terribleness of suffering. Secondly, whenever we ask why some action is right or wrong, we find we are led back to reasons that are not themselves about action but concern the good or evil of those affected by the action. For example, what makes the act of killing wrong is in the first place the evil of the death of the victim. Let us take these two reasons in turn to understand what is most central to ethics.

When we are confronted with pictures of genocide victims dug up, those of Srebrenica for example, we know 'those were people like us, and something terrible happened to them'. Our emotional reaction gives us an immediate insight into the violation and destruction of something of immense value, a human life. It is gross violations of the right to life that most immediately impose on us a sense of the objective inviolability of human worth. That is where we first understand how it is that ethics is objective – that good and evil matter in some absolute sense and that right and wrong cannot all be a matter of mere opinion or personal choice.

To be sceptical about something as ethically basic as the terribleness of the evil suffered by victims of genocide (or natural evil) would be not only a mistake but also an evil act against the victims. Raimond Gaita writes in *Good and Evil: An Absolute Conception*:

> The person who says that we must consider the argument that morality may be an illusion says that we must in intellectual conscience question, and if the argument goes a certain way have the courage to deny, the reality of the evil which at other times we think it our duty to remember. Of course no one is really serious about it. But who would confidently say that such frivolity is without consequence?[1]

No doubt objectivity can be found also in the demandingness of ethical commands, or in the absoluteness of rights and the nobility of virtues. But those do not have the same immediacy or urgency. It might be almost conceivable to suspect that absoluteness claims for commands or rights are a production of self-righteous fanatics certain of their own rectitude. But not evils. As Roy Holland puts the question 'When you learned of those monstrosities were you not horrified and did you not absolutely condemn them?'[2]

The worth of persons and the sense of tragedy about what happens to them is not the usual place to start in ethics. The range of topics commonly spoken of in ethics includes rules of right and wrong action, virtues, values, consequences, rights, obligations, dilemmas and so on. But when we look below the surface of each of them, we find that they lead back to the worth of persons and that they are unintelligible without that concept.

One way to recognize that there must be something ethically more basic than action is to consider how we think about whether rules should have exceptions. A list of rules cannot be the ultimate answer to basic ethical questions. We can always ask whether a rule is right – or not right, or right generally but with exceptions. We often do ask that question, and we have resources to answer it. For example, we ask about the rule 'Thou shalt not kill': could killing in self-defence be allowed? Surely protecting myself is good, since my being killed is bad. The *reason* for the rule 'Thou shalt not kill' is that being killed is bad, and that is not about rules of action, it is about the evil of being dead. The very reason for the rule implies that it could have an exception in the case of self-defence. Whatever we decide in the end about the allowability of self-defence, the debate about it forces us to think about the reasons that lie behind the rules. Those reasons must be more ethically basic than the rules themselves. They must involve the properties of persons – those properties that give persons worth.

What Gives Persons Their Worth?

If the worth of persons is to be founded in reality, it should be possible to state what it is about persons that gives them that worth. Their rationality, for example, seems relevant to worth, but not their size. We will first describe those properties unique to persons that are the foundation of worth, then explain the supervenience relation that holds of necessity between those (natural) properties such as rationality and the (ethical) property of worth.

A classic answer to the question, 'What gives humans worth?' is 'rationality'. 'Man's excellence,' says Saint Augustine, 'consists in the fact that God made him to His own image by giving him an intellectual soul *(mentem intellectualem)* which raises him above the beasts of the field"[3] A person is an 'individual substance of a rational nature' according to Boethius.[4] Kant (though his view of rationality is more inclusive than the purely intellectual) says '*rational nature exists as an end in itself.* In this way the human being necessarily conceives its own existence . . . In this way, however, every other rational being conceives its existence owing to just the same rational ground which also holds for me'.[5]

One reason why this answer is *prima facie* attractive is that it is the most obvious difference between ourselves and animals. The worth of animals, on the face of it, is less than ours, and that appears due to their lack of a full version of rationality since they resemble us closely in so many other respects. Conversely too, we value especially those animals like cats which are highly cognitive and inquisitive, and so have a trace of some kind of rationality. That is a large part of what makes them attractive as companion animals. If animals really did have the rationality attributed to them in anthropomorphic animal tales, we would not eat them and would treat them as real colleagues.

Intellectual rationality which is, as far as we know, unique to humans and possibly gods, is distinguished from the kind of 'intelligence' that can be handed over to Artificial Intelligence or found in animals. The major advances in Artificial Intelligence and the study of animal cognition in the last sixty years have shown the extraordinary powers of both but at the same time the large gap between the complex 'adaptive behaviour', calculation and search through spaces of possibilities that artificial and animal intelligence is capable of, and the uniquely human ability to genuinely understand. The sort of pure understanding of such truths as why 2×3 must equal 3×2, as described in Chapters 2 and 3 above, is resistant to Artificial Intelligence methods and there is no plan on how to start imitating it.

But even understanding rationality as widely as reasonable (but not so widely as to render it vacuous), the human person restricted to its rationality is autistic and not sufficient in itself to be the whole basis of the worth of persons. The first prominent thing it leaves out is the core of truth in Kant's dictum that the only thing good unqualifiedly is the good will. A good will is not a kind of rationality. It is a commitment. The human ability to choose actions for reasons is what makes such commitment possible. And as literary and psychiatric people have said at length, an exclusive focus on rationality omits the crucial emotional aspects of humanity that are central to what it means to be human. The whole emotional structure of humans, with their ability to give and receive love, is central to what makes them valuable.

Consciousness, too, is not exactly rationality but is essential to being human, as expressed in Descartes' dictum 'I think, therefore I am'. The ability to suffer, which we share with animals, and which gives them some moral standing, is a matter of consciousness, as is the ability to experience the best of human life, such as beauty, understanding, and pleasure.

Those suggested bases of the worth or persons, such as rationality, the capacity for free action, emotional structure, and consciousness are properties of humans, that is, universals or repeatables that humans can have in common. It has been argued that that approach leaves out something important: the individuality of persons. Does the search for *properties* as a ground of ethics necessarily miss the individuality or uniqueness of persons? Perhaps properly human worth does not supervene on properties shareable by humans at all (or not wholly so), but on individuality – on *who* a person is rather than *what* a person is.[6]

What 'individuality' means is not easily explained in words, since words deal in generalities. Philosophers deal, typically, in the abstract and repeatable. Novelists, on the other hand, even when representing figures who are in some sense types, do so through the individuality of the characters and their histories. Historians and especially biographers attempt something similar with real, particular, persons. So, we need to look more to literature and biography than to philosophy to understand the relation of individuality to worth. Those modes of writing are right to emphasize individuality as crucial to being human.

In any case, our common human experience, whether or not philosophically reflective, makes us well aware of the range of what it is that makes humans matter. When we see people losing those properties, for example to dementia, we have no need for an explanation of what a tragedy that is.

Supervenience of Worth on Non-moral Properties: Crossing the Is-Ought Gap

The nature of the relation between ethical worth and its grounding properties, such as rationality, is a difficult one. *Prima facie*, potential grounding properties like rationality (agency, individuality . . .) are specifiable without explicit reference to anything ethical (evaluative, normative . . .) while the worth that arises from them is paradigmatically ethical. Yet rationality (etc.) appears to be wholly sufficient to ground worth: once a person (or potentially an alien intelligence) has that combination of properties, there seems to be nothing further needed for what happens to it to matter morally. That appears to conflict with Humean intuitions about the is-ought or fact-value distinction, according to which an unbridgeable gap lies between the two.

Supposing we can identify some (metaphysical) properties of humans that ground or account for their worth, what exactly is the relation of worth to the grounding properties? What does the supervenience (in philosophical jargon) of worth on those properties mean? The supervenience or grounding relation that holds between rationality (etc) and worth will have to be a metaphysically powerful one. Though to an extent *sui generis*, we can hope to partly understand it through a comparison with a number of other, non-moral, relations of supervenience, especially that between logical relations and the normativity of reasons for belief.

So let us compare ethical supervenience with some examples of supervenience which are better understood – if not perfectly, then at least without the apparently special problems of ethical supervenience. Some possible comparators are

- The supervenience of (spatial) symmetry on the position of parts
- The supervenience of the meaning of words on spoken and written marks
- The supervenience of chess strategy on the rules of chess
- The supervenience of algorithms on their software implementations
- The supervenience of the betweenness relations among colours on the colours themselves
- The supervenience of the singleton set {Socrates} on Socrates
- The supervenience of a statue on its shaped material (the relation of material constitution)
- The supervenience of the rationality of belief on logical relations.

In each of these cases, the supervenient entities are substantially different in kind from the supervenience base. Yet we can understand – in principle fully and purely by thinking – how the base necessarily gives rise to the supervenient entities or properties. The necessity in every case is absolute, of the sort described in earlier chapters. We can see also how these entities/properties are therefore not 'queer' in the way that John Mackie alleges that ethical properties are too queer to fit in the scientific picture of nature,[7] nor are they supernatural. The supervenience of moral worth on the natural properties of persons is fundamentally the same as these kinds of supervenience. These cases provide models for the supervenience of moral worth and show there is nothing misconceived in the notion of supervenience itself.

The closest parallel is the last one, the supervenience of right belief on good reasons, which is (up to a point) well understood. The parallel with belief and reasons is a good pick because the space of reasons and the moral space both contrast with the space of natural causes in the same way. The base in each case consists of observable scientific properties, of brains and societies going about their causal business and issuing moral (respectively rational) conclusions, while the superstructure of absolute demands of morality (respectively rationality) appear *prima facie* to float free in a Platonic realm which imposes objective standards on the outputs of the causal processes. Whatever is, is caused, but 'Whatever is, is right' is equally false in the moral and logical worlds.

To parallel worth we would want something 'not about us', but reasons in the absolute and external sense of pure logic. As described in Chapters 4 and 5, there are such reasons, existing in logical space independent of our desires and interests – logical facts such as 'the evidence in court did not reach proof beyond reasonable doubt of the defendant's guilt', 'modus ponens is valid', 'the ontological argument is worthless'. 'Fermat's Last Theorem is proved'. They are authoritative or prescriptive in that for thinking to be right it must conform to them, but that is secondary to their obtaining as (necessary) facts in the logical realm.

The notion of the supervenience of worth on non-moral properties such as rationality is thus able to bridge Hume's 'is-ought' gap between the natural and the ethical. As he stated it, the gap was between matters of fact and oughts, and worth is not an ought (that is, a statement of an action being right) but a property of entities such as humans. Nevertheless, there is something correct, as has been widely felt, about Hume's assertion of some gap between 'relations of objects' or the purely scientific properties of the world, and ethical properties of any kind. The nature of that gap is not very clear. However, it is clear that the property of rationality lies on one side of the gap, the scientific side, and worth on the other. Piling up scientific facts about an object fails to include any facts (if facts they are) about its moral worth. Even if it is maintained that moral facts somehow supervene on nonmoral facts – for example, if it is maintained that anything with all human nonmoral properties, such as rationality and consciousness, is necessarily worth as much as a human – the moral facts are not found in the nonmoral facts. The gap requires some explanatory bridge. Supervenience is the bridge across that gap.

The parallel with the case of causes and reasons is again helpful, because that case lacks the special difficulties that are felt to accrue to the moral case. There exists a causes-reasons gap, similar to the is-ought gap. No amount of scientific facts about the causes of belief (such as Darwinian or neurological accounts of why we believe one thing or another) constitutes any reason for those beliefs. The evidence, in say a court of law, is a sum of statements of fact, and the reasonableness or otherwise of the verdict supervenes on those (claimed) facts – with the evidence as it is, only one verdict is rationally justified. But the causes of the jury's verdict in their upbringing and neurology are not part of that supervenience. We may feel we do not fully grasp how reasons supervene on facts, but that cannot be because of any 'queerness' of the logical. The logical will have to fit into the scientific world picture somehow, since science depends on – in fact largely consists of – logical inferences. We understand the supervenience of reasons on facts sufficiently to provide an encouraging parallel for the ability of supervenience theory to bridge Hume's gap.

We have spoken as if supervenience is a heavy-duty and somewhat unfamiliar metaphysical notion. That is so, but at the same time it is implicit in how people commonly think about ethics, even those who claim not to accept any such notions. In a well-known article on 'The stupidity of dignity', Steven Pinker rejects the concept of human dignity (the more common name for worth in religious and human rights traditions), on the ground that it is 'a squishy, subjective notion, hardly up to the heavyweight moral demands assigned to it.' But then he adds that 'bioethics has done just fine with the principle of personal autonomy – the idea that, because all humans have the same minimum capacity to suffer, prosper, reason, and choose, no human has the right to impinge on the life, body, or freedom of another.'[8] Anyone who understands his 'because' has a basic grasp of supervenience, and anyone who accepts that humans have rights because of a 'minimum capacity to suffer, prosper, reason, and choose' understands something of the moral bases on which worth supervenes.

How Worth Gives Rise to Moral Obligation

Hume's talk of an is-ought gap, though pointing to a genuine problem, conflated different gaps. It is important to distinguish between the natural properties-worth gap (argued above to be bridged by the notion of supervenience) and the worth-obligation gap. (And those need to be distinguished from the further gap between obligation and motivation, which will not be discussed here.)

Worth is an intrinsic property of certain entities, such as humans. Moral obligation is quite another kind of thing. It is a relation between an actor and a potential action (or attitude) of that actor. The worth of humans is a (moral) fact about them, but my obligation to assist or respect them is a relation of me to them. If intrinsic worth is somehow to give rise to moral obligation, some substantial story needs to be told to fill the gap between worth and obligation. As moral error theorists point out, there is something mysterious about how facts – any sort of facts – could have or imply 'requiredness'. 'The problem . . . is precisely to explain how there can be facts that *in themselves*, i.e. irrespectively of the desires, aims, or roles of human beings and other agents, require, or *count in favour of*, certain forms of behaviour.'[9]

Yet at the same time, the connection between worth and obligatoriness is experienced by normally functioning humans as immediate and obvious. If I see someone fall in the river and there is no-one else to rescue them, my recognition of the impending harm to them generates an obligation on me to help them (if I can) and I should be happy to shoulder the burden (and feel no need for or benefit from a command to fill the gap). What could there be to deliberate about ethically, given the immediate need and my ability to do something about it?

It will help to consider again a parallel case which, though not exactly clear, lacks the special difficulties of the moral case. It is the same parallel case that helped understand the supervenience of worth on natural qualities, namely, the parallel with logical reasons for belief. There is another kind of 'should', the ought of logic. Why should one believe a proposition that is implied by (or highly probable given) one's total evidence? The essential answer to the question 'Why believe what is implied by or made probable by one's evidence' is just 'Because it is implied or probable'. Why not believe that the earth is flat? Because the evidence is against it. That is the complete justification and nothing can add to it.[10]

It is similar with worth and its relation to moral obligation. If someone falls in the river near me, it is my responsibility to help him if he appears to need it. That is just because he is another human being in extreme need and I have the ability to save him. His right to be saved and my duty to save him follow from his need and my ability. A prospective injury which I could easily prevent is a harm to something of great worth. If I don't act, I have (partially) caused that harm. It directly imposes on me a moral obligation to prevent it, and nothing can add to that – not any advantage to me, or sympathy with the victim, or enhancement of my virtue, or command of God or human law, or theory about rationality or universalizability.

Knowing the Good

As in the case of mathematics, ontology comes first, epistemology second. But once the nature of the good has been established, some story must be told about how it is known. In particular, if humans have worth, how is that known? As with mathematics, knowing necessities requires a different style from knowing empirical facts.

We could proceed in one of two ways. First, we could consider and adapt reasonable suggestions that have been made about ways of knowing objective moral facts in general. Possibilities include:

- A priori intuition, as of basic mathematical truths
- Moral perception
- Emotional reactions such as empathy and gut feelings on witnessing atrocities
- Moral development by understanding injustice to ourselves and applying a symmetry argument
- Love as revelatory of the dignity of others
- A transcendental argument that anything capable of understanding worth must itself be valuable
- Divine revelation that all humans are made 'in the image and likeness of God'.

The second alternative is to proceed along the path just outlined for the metaphysics of worth: explain knowledge of the 'sub-moral' bases like rationality and individuality, then of the supervenience of worth on them. If we were optimistic, we might hope that pursuing both paths would lead to their meeting in the middle. In fact, they do.

Firstly, we know well the bases of worth. The properties of humans on which worth supervenes – rationality, consciousness, the unified self, the diversified self, the rational will with freedom, love and the emotions, individuality – are natural in the sense of not being explicitly ethical. The ethical supervenes on them, rather than being identical with them. But they are not exactly scientific properties like mass, length and charge either. The 'detectors' of those properties cannot be purchased from a supplier of lab equipment but are more intimately related to how it is to be human. To understand what it is like to be another human, with a unique life history and experience, point of view and emotions, requires a kind of imaginative sympathy that can be objectively right or wrong, but which contrasts with the method of the natural sciences.

Rationality and consciousness are easy. At least knowing one's own is. 'I think, therefore I am rational' and 'I think, therefore I am conscious' are as incontrovertible as 'I think, therefore I am' and for the same reason. If you are reading this book (which you are) you are rational – if you weren't rational, you couldn't read it. Knowing our own emotions, individuality and so on takes some more maturity, but, with help from the social world, it is possible. Non-autistic humans have ways of developing a 'theory of other minds', which enables them, slowly, to appreciate that other humans have those qualities too.

Given that we know well the bases of human worth, how then do we know the supervenience of moral worth on them? Normally, no further step is needed.

Once we know rationality, emotions, the uniqueness of (actual, individual) persons, it 'goes without saying' that we have understood their worth. Given the intimacy of the supervenience relation, something would be wrong if the step was not made. It is certainly possible to go wrong, as in cases of psychopathy or of self-hate arising from early trauma but, in those cases, we look for psychiatric causes of a failure of normal understanding.

We can, however, usefully compare the case to other cases of supervenience, where there is something to be known but the intimacy of the supervenience relation makes it hard to say anything contentful about the intuition needed to know it. Take again the supervenience of the rationality of belief on logical relations. How, exactly, do we know that if our evidence implies a proposition, it is reasonable – we logically ought – to believe it? Logical relations are one thing, our beliefs another: how do we know the connection? It is certainly possible, indeed only too common, to have a kind of 'logical psychopathy' that believes things based on blind faith or wishful thinking or righteous indignation instead of logical relations. It is not just Lewis Carroll's Red Queen who has believed as many as six impossible things before breakfast. So, there is something non-trivial to be known. How, then, do we know that if I know *p* and know that *p* implies *q*, then I ought to, or would be well-advised to, believe *q*? (In general, logical relations are normative for belief.) It is not a matter of observation, and any further reasoning for it would be circular.[11] We are reduced to invoking 'intuition' or understanding, as in mathematics. The understanding is a very direct mode of knowing. It is most familiar in mathematics and logic, where its direct deliverances (such as understanding why $2 \times 3 = 3 \times 2$, or why 'the set of points equidistant from one point' defines a circle, or why affirming the consequent is invalid) form the basic truths of those disciplines.

Given the abstractness of matters like supervenience and understanding, it is fortunate that there are also available more down-to-earth and familiar roads to basic ethical knowledge. They involve empathy, love, and similar emotional reactions to the ethical nature of persons and demands for justice. We will take just two examples.

Both are found in infants. Their perception of justice was emphasized by Piaget's classic work on children aged about 5 to 7 years. (As Dickens says, 'In the little world in which children have their existence . . . there is nothing so finely perceived and so finely felt, as injustice.'[12]) Children (at least those living in an appropriately supportive culture) are able to develop concepts of fairness, by recognizing that other minds are not relevantly morally different to their own, when it comes to getting what they deserve.[13] Piaget describes how very young children believe that what is right is simply what is approved by adult authority. But at a later stage they develop a sense of fairness based on a sense of equality. Later research on the moral development of infants has tended to show that what Piaget was looking at was already a considerably developed stage, and there are more basic underpinnings of ethical concern visible at ages as far back as one or two. Those earlier stages do not involve just acceptance of parental rules, as Piaget thought, but a clear feeling of natural sympathy for others.[14]

At a later stage in life, our ability to acquire moral knowledge by immediate emotional empathy with other humans can be extended by serious novels – for example, when Pasternak in *Doctor Zhivago* has the fully developed character of Lara disappear into the Gulag, it is the empathy the reader has developed with the character that points up the moral horror of a political system that treats people as anonymous vermin. A fundamental demand of humans to be recognized as human by others is one of the 'needs of the soul', in Simone Weil's words;[15] it is prior (in knowledge) to any speculations about what features of human nature may generate it, or any identification of rights. And, without at least some of that initial emotional attunement to the irreducible worth of humans, there can be no meaning to discussions of human nature or rights.

Worth of the non-human

Although most of our moral dealings are with humans, some are with other entities, such as animals we might eat, ecosystems we may destroy, and gods we may worship and obey. Something must be said about whether those interactions depend on their having any degree of worth.

The properties that have been argued to confer worth, such as consciousness, emotions, and individuality, and possibly rationality, can come in degrees. If animals possess them to some degree, then they have a degree of worth. Some more and some less.

In the area of rationality, argued earlier to be especially important for human worth, it is especially important to avoid either minimizing or exaggerating the resemblances between humans and higher animals. Everyone knows from meeting higher animals that they are highly cognitive, for example, they are very smart at navigating. Certainly, modern research has shown that the cognition of apes, dolphins and octopuses is considerably more sophisticated than was once assumed. But 'cognition' is not exactly rationality. It remains an uphill job to bridge more than a little of the gap. It is surprising that humans share 98% of their DNA with chimpanzees. But the chimps themselves are not surprised by that, and they cannot be surprised by it. Their mental world just does not have that whole realm of theoretical understanding, abstraction, and reflection on concepts.[16] We can discuss what it is like to be them, up to a point, but they cannot wonder what it is like to be us.

Similar remarks apply to the 'rational will' that Kant thought crucial to human dignity. Again, one should neither underplay nor exaggerate human-animal parallels. Cats can in a way plan and decide. A cat can plan a route over fences to arrive on the back doorstep at dinnertime. That makes cats suitable as companions and objects of concern. But that is not the same as making decisions in the sense of acting on understood reasons, nor integrating them into a life narrative, nor being capable of explaining why they did it. In the same way, some of the other bases for worth suggested for humans, like unity of consciousness, emotions, and individuality, are shared to a certain degree by animals, but far from completely.

It follows that animals can correctly be said to have at least some rights, for the same reason as humans do – their worth puts a moral limit on what can be done to them. But they do not have rights or worth equal to humans.

The non-experiencing parts of the universe, from gas clouds up to plants and sponges, cannot have the sort of worth that humans and higher animals have. They do not have any degree of rationality, consciousness, individuality or any of the other features argued earlier to be the bases of worth. They do not suffer. They have no literal welfare or interests. Some of 'them', such as corals or ecosystems, are hardly individual entities at all. If they have any worth – inherent worth over and above their usefulness to experiencing beings – it will have to be an alien sort of worth.

Parts of creation like rainforests and artifacts do have an entirely different but still genuine and objective kind of value. It would be better called aesthetic value than ethical, were it not for the almost irredeemably subjectivist connotations of 'aesthetic'. Older, objectivist views of the aesthetic will need to be recovered and adapted to some contemporary environmentalist thought to explain what is meant.

The notion of vandalism provides an entrée into the worth of non-sentient things, natural or artifacts, in the same way as tragedy does for the worth of humans and cruelty does for the worth of animals. Those concepts are about human response, but they make sense because they are correct responses to losses independent of us.

Vandalism, the 'ruthless destruction or spoiling of anything beautiful or venerable' (OED) has an aesthetic aspect to its meaning. But that means the aesthetic in an objective sense, not as a matter of mere taste nor as descriptive of what is in the 'eye of the beholder'. If we say that our attitudes to sentient life forms are moral whereas our attitude to the rest of the universe is aesthetic, that does not imply less objectivity in respect of the latter.

If we accept that things or systems (like ecosystems) can be the bearers of aesthetic worth, it remains to ask what properties of systems, or parted wholes, could be the bases of that worth. What properties can systems have, on which 'aesthetic' worth can supervene?

The classic statement of the bases of aesthetic worth, that of Aquinas, is relevant although as it stands it is applicable only to the beauty of individual things, rather than, for example, ecosystems. He writes:

> Three things are necessary for beauty: first, integrity or perfection, for things that are lacking in something are for this reason ugly; also due proportion or consonance; and again, clarity, for we call things beautiful when they are brightly coloured.[17]

Those properties are all, in Aquinas's view, inherent in the things themselves and not at all in the eye of the beholder. The complexity, symmetry and similar properties (according to one list 'complexity, stability, diversity, connectedness, variety, subtlety, intricacy, ingenuity in some suitably non-intentional sense, fragility, harmony'[18]) that are normally praised as the good-making features of ecosystems or landscapes are the same kind of properties as Aquinas's 'proportion or consonance'. Those are strictly mathematical properties – concerned with the same and the different, ratios, symmetry, equality of parts. They are objective mathematical properties of systems that are parts of the natural world. Aesthetic worth supervenes on them, with the result that we have a moral obligation to appreciate and preserve their beauty.

God, as conceived by classical theism, has another kind of worth. The claim of Genesis that humans are 'made in the image of God' suggests it would be a superlative version of our worth rather than alien like the worth of rainforests. If such a being exists, worship and obedience would be natural reactions to our knowledge of it.

Conclusion

The Victorians came to find personal theisms and their dogmas unconvincing, yet felt the absolute force of moral claims. Some found the conflict particularly oppressive, as recorded in Frederic W.H. Myers' recollection of a conversation with George Eliot in 1873:

> I remember how, at Cambridge, I walked with her once in the Fellows' Garden of Trinity, on an evening of rainy May; and she, stirred somewhat beyond her wont, and taking as her text the three words which have been used so often as the inspiring trumpet-calls of men—the words *God, Immortality, Duty*—pronounced, with terrible earnestness, how inconceivable was the *first,* how unbelievable the *second,* and yet how peremptory and absolute the *third.* Never perhaps, have sterner accents affirmed the sovereignty of impersonal and unrecompensing Law. I listened, and night fell; her grave, majestic countenance turned toward me like a sibyl's in the gloom; it was as though she withdrew from my grasp, one by one, the two scrolls of promise, and left me the third scroll only, awful with inevitable fates. And when we stood at length and parted amid that columnar circuit of the forest trees, beneath the last twilight of starless skies, I seemed to be gazing, like Titus at Jerusalem, on vacant seats and empty halls—on a sanctuary with no Presence to hallow it, and heaven left lonely of a God.[19]

In another celebrated Victorian dictum, 'Opinions alter, manners change, creeds rise and fall, but the moral law is written on the tablets of eternity.'[20]

Twentieth-century writers were more inclined to celebrate their 'liberation' from moral constraints. But those who did retain a sense of the objectivity of morality found it hard to explain where such objectivity could come from, given its omission from the scientific world picture.

The foundation of ethics in the worth of persons answers that question and restores the parallel between the absolute necessity of the truths of ethics and the absolute necessity of the truths of mathematics and logic.

Notes

1 Raimond Gaita, *Good and Evil: An Absolute Conception* (2nd edn, Abingdon: Routledge, 2004), 315.

2 Roy Holland, 'Absolute Ethics, Mathematics and the Impossibility of Politics', in *Against Empiricism* (Oxford: Blackwell, 1980), 126–42, at 129.

3 Augustine, *De Genesi ad Litteram*, bk 6 ch. 12, quoted in Thomas Aquinas *Summa Theologiae* I q. 93 art. 2; discussion in Christian Tornau, 'Augustine of Hippo', *Stanford Encyclopedia of Philosophy* (2019/2024), section 6.2.

4 Boethius, *Liber de Persona et Duabus Naturis contra Eutychen et Nestorium*, ch. 2.

5 Kant, *Groundwork for the Metaphysics of Morals*, 4:428–9.

6 Hannah Arendt, *The Human Condition* (2nd edn, Chicago: University of Chicago Press, 1958), 242.
7 John Mackie, 'A Refutation of Morals', *Australasian Journal of Philosophy* 24 (1946): 77–90.
8 Steven Pinker, 'The Stupidity of Dignity', *New Republic*, 28 May 2008.
9 Jonas Olson, 'In Defense of Moral Error Theory', in *New Waves in Metaethics*, ed. M. Brady (Basingstoke: Palgrave Macmillan, 2010), 7.
10 David Stove, 'Why Should Probability Be the Guide of Life?', in *Hume: A Re-Evaluation*, ed. D. W. Livingston and D. T. King (New York: Fordham University Press, 1976), 50–68.
11 Susan Haack, 'The Justification of Deduction', *Mind* 85 (1976): 112–19.
12 Charles Dickens, *Great Expectations*, ch. 8.
13 Jean Piaget *et al*, *The Moral Judgment of the Child* (London: Routledge & Kegan Paul, 1932); Charles C. Helwig, 'Children's Conceptions of Fair Government and Freedom of Speech', *Child Development* 69 (1998): 518–31.
14 Alison Gopnik, *The Philosophical Baby: What Children's Minds Tell Us about Truth, Love & the Meaning of Life* (London: Bodley Head, 2009), 204.
15 Simone Weil, *The Need for Roots* (London: Routledge & Kegan Paul, 1952), 3.
16 John Haldane, 'Rational and Other Animals', *Royal Institute of Philosophy Supplement* 41 (1996): 17–28.
17 Aquinas, *Summa Theologiae*, I q. 39 art. 8; Umberto Eco, *The Aesthetics of Thomas Aquinas* (Cambridge, MA: Harvard University Press, 1988), 65; James Franklin, 'The Realism of Aquinas: The True, the Good, the Beautiful', *The Sydney Realist* no. 38 (Sept 2019): 9–13.
18 Robert Elliot, 'Environmental Degradation, Vandalism and the Aesthetic Object Argument', *Australasian Journal of Philosophy* 67 (1989): 191–204, at 193.
19 Frederic W.H. Myers, 'George Eliot', *Century Magazine*, Nov 1881 (http://www.bartleby.com/309/1001.html), discussed in William J. Wainwright, *Religion and Morality* (Aldershot: Ashgate, 2005), 3.
20 James A. Froude, 'The Science of History' (1864), in *Representative Essays Selected from the Series of 'Prose Masterpieces from the Modern Essayist.'*, ed. George Haven Putnam (New York: G. P. Putnam's Sons, 1885), 250–84, at 274.

Part Six

History: The Rise and Fall of Absolute Necessities

11

An alternative, aristotelian, history of science

While the philosophy of absolute necessities, like any other philosophical problem, can be studied quite apart from history, there are advantages for both philosophy and history in connecting the two. Taking a philosophical perspective allows a thread to be identified in the history of ideas that might not be visible otherwise, organizing what seemed to be unconnected fragments into a coherent history of a theme. On the other hand, understanding the history of the philosophical problem brings centuries of intelligent thought to bear on it, showing how the problem developed in the minds of those who moved it forward.

It might be most natural to start with the history of mathematics, since mathematics is the home ground of absolute necessities. The history of mathematics is a well-researched area but, unfortunately, has often been presented from a Platonist point of view, with problems originally 'motivated' by physics being then studied as abstract theorems about sets, functions, vector spaces and so on. Later in the chapter we will consider some history of mathematics since the Scientific Revolution from a more applied perspective, but we will begin instead with the history of logical probability. That is an area where absolute necessities appear much more commonly in ordinary reasoning than do complex mathematics. Ethics, too, was an applicable science where necessities were recognized at an early date.

Medieval Law and the Origins of Probability Theory

As we saw in several previous chapters, a conceptually based science of necessities was bequeathed to the later world by Aristotle and Euclid. But science by conceptual analysis and deduction was by no means confined to mathematics, or even to what was later called (in English) science. Where (geometrical) space proved amenable to logical deductions, non-deductive reasoning – the logical probability studied in Chapter 5 – found its natural place in law, in particular the law of evidence.

Law deals with evidence, hence with the strength and conflict of evidence. The two great systematizations of law of late antiquity, the Jewish *Talmud* and the Roman *Corpus of Civil Law*, both insisted on very high standards of proof for criminal conviction but recognized that the standard was short of absolute certainty. In Jewish law, conviction required two witnesses, who had to give direct evidence that was not 'based on conjecture' but direct. Roman law held that conviction may not be 'on suspicion', for 'it is better to permit the crime of a guilty person to go unpunished than to condemn one who is innocent'. The charge must be 'supported by suitable witnesses, or established by the most open documents, or brought to proof by signs that are indubitable or clearer than light'.[1] Torture was a feature of Roman law; judges investigating public crimes were not, however, to begin with torture, but to use whatever 'likely and probable arguments' (*argumentis verisimilibus probabilibusque*) were already available. The word 'likely' (*verisimilis*), found earlier in Cicero, was used a number of times in the *Corpus of Civil Law* with very much the common meaning it has in modern English, in examples like 'it is very likely (*verosimilius*) that the testator had intended rather to point out to his heirs where they could readily obtain forty aurei . . . than to have inserted a condition in a trust'.[2] Both Jewish and Roman law included a concept of 'presumption': a proposition taken to be true until the opposite is proved; some but not all of these are likely but uncertain generalizations.[3]

Ancient Roman law was rather untheoretical, but when the *Corpus* was rediscovered in Italy in the late eleventh century and began to inform Western medieval law, its commentators sought to understand the principles and justifications of the Roman rules. In commenting on evidence, they came to realize that there must be pieces of evidence and presumptions of different strengths. The school of Glossators, of the late twelfth century, united several scattered remarks of Roman law under the classification half-proof (*semiplena probatio*): 'It would seem that . . . either the plaintiff proves, or not . . . I reply that although according to Aristotle it would seem to be an exhaustive division, it is not so according to the laws. There is a medium, namely, half-proof. Say therefore that in such a case proof has been less than full. Then the judge gives the right to complete the case by oath to the plaintiff . . . Full proof is by two witnesses, therefore half proof by one.'[4] There was never any finer grading attempted, such as quarter-proofs. However, a qualitative grading of the strength of presumptions became standard. In a decree of Pope Innocent III in 1209, the question concerns someone in doubt as to whether there is some legal impediment to his marriage and hence wondering if sex with his/her spouse is permitted. It depends on the strength of the evidence:

> It should be distinguished, whether the spouse knows for certain the impediment to the marriage, and then he may not engage in carnal intercourse . . . or whether he does not know the impediment for certain, but only believes it . . . In the second case, we distinguish, whether his conscience is thus from a light and rash belief, or a probable and discreet one (*ex credulitate levi et temeraria, an probabili et discreta*) . . . when his conscience presses his mind with a probable and discreet belief, but not an evident and manifest one, he may render the marriage debt, but ought not to demand it.[5]

The theory was elaborated by legal writers and applications made. Torture, reintroduced under the influence of Roman law, could be applied only when there was half-proof of guilt (to produce the other half). The smallest quantum of evidence considered was the 'indiciolum', a diminutive of 'indication' or 'sign': If in a murder case a witness testifies that the brother of the accused was an enemy of the deceased, 'an indication is certainly not thereby proved, although it makes an *indiciolum* which is very small, and of no strength'.[6] The theory of grades of evidence was laid out with many distinctions in the fourteenth-century commentary of Baldus de Ubaldis, which remained standard for centuries. Massive sixteenth-century expansions weighed down by many examples included Mascardus's *On Proofs* (1584) in three volumes and Menochius's *On Presumptions, Conjectures, Signs and Indications* (1587), in two.[7] Although England had a separate legal tradition from the Continent, Latin concepts such as violent presumptions and half-proofs made their way into seventeenth-century English law.

'Probability', in this evidential sense clarified by legal theory, penetrated other fields, such as the sceptical evaluation of historiographical evidence and, in Galileo, the weighing of evidence for the Copernican theory of the heavens.[8]

Probability in Ethics: Probabilism in Moral Theology, Aleatory Contracts

In medieval and early modern times, ethics, like law, was not as separate from what became known in English as 'science' as it later was. Much of the development of ideas on probability and risk took place in ethical contexts.

Before the word 'probability' became attached to stochastic processes like dice-throwing, it was best known in the context of evidence, especially in the doctrine of 'probabilism' in moral theology. The background was the medieval moral maxim that 'in cases of doubt (about what is morally right) the safer part is to be chosen'. That could lead to excessive scruples based on fanciful doubts, so the principle was gradually attenuated. In 1577 the Spanish Dominican Bartolomé de Medina set out the position of probabilism, which gradually became Catholic orthodoxy: that in a case of doubt on what course of action is allowed (such as a conflict between reasons and authorities), one might choose a probable course of action, even if the opposite course was more probable. A proposition was said to be 'probable' when it was supported by some sound reasons and/or substantial authorities. Heated debate between strict and lax interpretations continued up to Pascal's *Provincial Letters* of 1656–57, which had great popular success in accusing the Jesuits of lax moral opinions based on their probabilism.[9]

At the same time, moral theorists analysed the justice of aleatory contracts – contracts whose fulfillment depends on chance, such as insurance, life annuities, purchase of options, and games of chance. In all such cases, the payout that results depends on a chance event, such as the loss of a ship or the fall of dice, so the question arises: What is the just price for a contract before the event? The price depends on what we would call a probability (whether stochastic or epistemic). Theorists recognized that and discussed the factors affecting, for example, insurance rates. That was the background to the question Pascal and Femat discussed in founding the mathematical theory of probability in 1654: What is the just division of the stake in an interrupted game of chance?[10]

Science by Conceptual Analysis: The Genius of the Late Scholastics

The late scholastics – from the fourteenth to the seventeenth centuries – have suffered from a severe reputational problem. Attacked by every major thinker from Descartes to the Encyclopedists, they were a favourite target of abuse for Enlightenment rewritings of intellectual history, representing the darkness from which the Enlightened offered to rescue humanity.[11]

The scholastics adopted the Aristotelian ideal of the intellectual enterprise, as described in Aristotle's *Posterior Analytics* and implemented in Euclid's *Elements*: that certain knowledge could be gained by the intellect's reflection on experience and organized as a deductive structure following from simple axioms known with certainty. They added to that ideal a training in argument designed to reveal the true axioms and to refute errors, and an embodiment of the result in a textual tradition and an institution designed to carry the tradition forward from generation to generation – the university.

Not all branches of knowledge are well adapted to that mode of discovery and transmission. What is remarkable, as the late scholastics showed, is how many are. A central one is the analysis of continuity and change that formed the basis of the calculus developed in the seventeenth century. It proved essential to the work on physics of Newton and Leibniz and their successors, but it is not itself physics. It is an absolutely necessary mathematical science, accessible by pure thought.

The work of the fourteenth-century 'Merton School' and its French collaborators Jean Buridan (c 1300–59) and Nicole Oresme (c 1322–82) has been intensively studied and is well displayed in many selections in Grant's *Source Book in Medieval Science*.[12] It has been highlighted again in Hannam's book on the medieval roots of modern science, *God's Philosophers*.[13] However it has usually been thought of in the context of physics, and the scholastics have been regarded as 'forerunners of Galileo'.[14] That is true as far as it goes, but it distracts attention from their achievements in the pure analysis of continuity and change. Their discoveries are better explained in Boyer's *History of Calculus*, where they are recognized as an essential part of the process of the analysis of rates and change that was completed by the calculus of Newton and Leibniz.[15]

Mathematics is usually regarded as a model of absolute clarity of thought, achieved through the use of symbols, and a strange place for scholastics to be delving in. But a pre-requisite of most mathematics is analysis of concepts confusedly present in experience, and in this the scholastics were masters. Ancient mathematics, mainly mathematical astronomy, had some understanding of large-scale uniform motions and their combinations, but there was no understanding of small-scale motions. For example, there is no unit of speed (such as 'miles per hour') in ancient languages, no doubt largely due to the impossibility of measuring small units of time. The first thing needed to begin the study of motion (or any other rate of change) is to distinguish between uniform and non-uniform speeds – in mathematical terms, between a first and a second derivative. That distinction is a matter of pure conceptual analysis. The fourteenth-century calculators not only did that but also distinguished further between uniform and non-uniform accelerations. Then they showed a deep understanding of uniform acceleration by proving the 'Merton mean speed theorem': that a uniformly accelerated body travels the same distance as a body at uniform speed whose speed equals half the final speed of the accelerated body.[16] In the later language of calculus, that is to show that the integral of a constant function is a linear function.

It is true that the scholastics did not establish the main application of their analysis in physics, the fact that heavy bodies dropped near the earth's surface *are* uniformly accelerated. That is a difficult fact to discover, because the fall of such bodies is too fast to measure easily (without modern technology such as time-lapse photography). It was established with considerable experimental ingenuity by Galileo. Nevertheless, the suggestion that falling bodies *might* be uniformly accelerated was made by a scholastic whose work was known to Galileo, Domingo de Soto (1494–1560).[17] Indeed, careful studies have shown that the early Galileo had the mentality of a late scholastic himself.[18] For example, his graphical proof of the mean speed theorem is very close to Oresme's.[19] (Oresme also invented the concept of graphs, one of the few mathematical discoveries since ancient times that are familiar to every reader of newspapers, and one of the crucial tools for thinking about functions.)

Analysis of infinities and the continuum is also suited to the conceptual methods of the scholastics;[20] Albert of Saxony (c 1320–90) pairs off infinite sets 'in the imagination' in the same style as Cantor.[21]

It is true that the more down-to-earth empirical sciences were inappropriate for the scholastic method. Chemistry, pharmacology and metallurgy are examples: knowledge of the effects of chemicals and drugs and the strength of materials is so heavily empirical that looking into oneself and analysing concepts will not achieve much progress. It is no accident that these disciplines were undeveloped before very modern times.

On the other hand, the more conceptually-orientated of the social sciences – economics, psychology, political and legal theory and linguistics – were well-adapted to the scholastic method. It is a commonplace that those disciplines are still riven by disputes about theory in a way that the natural sciences are not. They are full of vigorous debates about what concepts should be used to organize and ask questions about the empirical data – even debates about whether any empirical data are free of theoretical perspectives. That is an indication that those fields are prime candidates for the conceptual analysis that was the strong suit of the late scholastics. We will survey briefly three disciplines by way of example: economics, psychology and linguistics.

In economic history, the idea that the serious development of theory began with the seventeenth-century Mercantilists (perhaps with Aristotle deserving some passing mention) was exploded by Schumpeter and others, with their examination of the scholastics of, especially, the sixteenth century,[22] but it can hardly be said to be common property even now. Concepts like demand, capital, labour, utility, scarcity and the cost of lost opportunity were developed at length by the scholastics, based on Aristotle's brief remarks. It is difficult to evaluate the contribution of the business community itself to these notions, for lack of evidence, but at present it seems that the explicit names and distinctions were due more to the scholastics' need for precision in ethical and legal discussions than to any strictly business requirements. In any case, the distinction between ethics and business was once not so rigidly drawn. Oresme's monetary theory is only one of the connections between the scholastics and more applied and mathematical investigations into business.[23]

Psychology, especially of the introspective kind, was also well-adapted to the scholastic method, although in this case the main developments came earlier. The general medieval concentration on the inner life encouraged by Augustine, the self-examination of enforced confession and Aristotle's *De anima* is reflected in the extensive attention given to the mind and soul in scholastic works from the time of Albert the Great and Aquinas, and the large number of texts on psychological subjects in the sixteenth and seventeenth centuries.[24]

One of the longest-running themes in the history of psychology is 'faculty psychology', as advanced by Avicenna and his Western scholastic followers from Aquinas on. Faculty psychology holds that human cognition is decomposable into largely distinct tasks like sensation, comparison of different sensory modalities, memory and so on, which are undertaken by different 'faculties' located in different regions of the brain. It was the natural way of organizing psychological data for many centuries,[25] and its (Latin) language is the way we still express our 'folk' psychological explanations of behaviour in terms of aspirations, sensations, imagination, actions, motives, emotions and so on.[26] As with the language of science, it is no accident that these words are Latin and the senses in which we use them are medieval more than classical. As a typical example, the *OED* traces the etymology of 'imagination' thus: in classical Latin, it means the action of forming mental images, but in medieval Latin and European vernaculars, it is in the first instance the faculty in which mental images are formed: in Old French c. 1174, Catalan and Italian late thirteenth-century, Spanish and Portuguese fourteenth century, English c. 1340.[27] The imagination was to prove a useful medium for thought experiments in science, notably in Galileo.[28]

While the approach was disparaged in the 1950s heyday of behaviourism, the revival of faculty psychology under various names has been one vigorous strand in more recent cognitive psychology.[29]

Grammar was a foundation stone of medieval education. From the twelfth century, the inherited texts were subjected to the same process of glossing and commentary as legal texts, in search of the philosophical principles underlying them.

Many of the concerns of modern semiotics on the different ways in which words can have meaning are visible in medieval debates about signs. For example, Augustine's conception of a sign as something that is perceived and suggests something else to some mind raises questions about signs of the non-existent, signs not interpreted by any mind, and about the sameness of 'mental words' in the minds of speakers of different languages.[30]

Vocabulary is one of the great hidden successes of scholasticism. The present-day international language, in which this book is written, is a blend of Old English, Norman French and scholastic Latin; and it is the vast inheritance of scholastic vocabulary, mostly still carrying its medieval meaning, that bears the weight of academic discussion on all subjects.[31] Consider the current section in general, or the present sentence in particular, or examine adjacent portions of the text, for a variety of natural and artificial examples of scholastic abstract vocabulary. That vocabulary still carries the deposit of Aristotelian thinking about necessities, if we care to look.

The Aristotelian Scientific Revolution

We now return to more mathematical material and consider the story of the discovery of mathematical necessities in physical reality from the time of the Scientific Revolution to the present.

The prestige of the Aristotelian-Euclidean model of science was highest in early modern times, after the vast extension of the realm of mathematics in the Scientific Revolution but before the discovery of non-Euclidean geometries and the foundational crises in mathematics of the late nineteenth century. When the Jesuit missionaries chose Euclid as the first book to be translated into Chinese[32] and when Spinoza offered to exhibit ethics as a system demonstrated *more geometrico* from definitions and simple self-evident axioms, they paid tribute to the place of the Euclidean model at the summit of intellectual achievement.

Seventeenth-century thinkers who applied mathematics to physics – Stevin, Galileo, Descartes, Pascal, Huygens, Barrow, Newton, Leibniz, to name only the most prominent – were captivated by a model of applied mathematics, apparently realized in Euclid's *Optics* and Archimedes' mechanics, according to which pure thought could establish principles for empirical reality. They were particularly impressed by Archimedes' demonstration of the law of the lever from pure symmetry considerations, as described in Chapter 9 above. Their ambition stood at the opposite extreme from the Baconian style of 'experimental philosophy' which recommended generalization from carefully collected facts.[33]

Their hopes were not fully realized, because of such awkward facts as the contingency of the constant of gravitational attraction. But their vast optimism concerning the power of mathematical reasoning and the possibilities of understanding reality through pure symmetry arguments proved astonishingly successful and created modern physics.

A well-known example at the very beginning of the Scientific Revolution, where pure symmetry considerations led to a surprising and apparently empirical result, is Stevin's 'wreath of spheres'. Stevin printed the diagram as the title page of his 1586 *Elements of the Art of Weighing*, a work which imitates and extends Archimedes' deductive structure.[34]

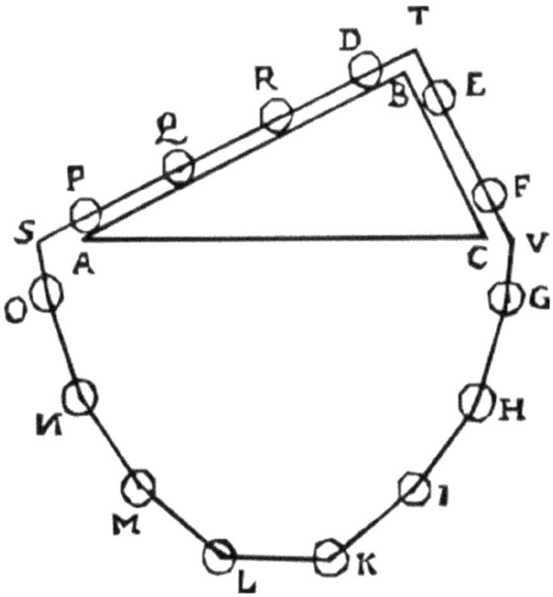

Figure 11.1 Stevin's 'wreath of spheres' showing how weights on inclined planes balance each other according to the angle of inclination.[35]

The text which decorates it says, approximately, 'Wonderful, but not incomprehensible'. That is correct. It is clear that the circle of balls does not tend to rotate either clockwise or anti-clockwise (that would be perpetual motion and, in any case, there is no reason to prefer clockwise to anti-clockwise rotation).

But the balls hanging below the horizontal line are in equilibrium. One could cut them in the middle, and allow them to hang down, or even remove them altogether, without disturbing the balls resting on the two inclined planes. These upper balls, then, are at rest, in balance: the many on the lightly inclined plane balancing the few on the steeper plane. The numbers of balls on each side of the apex are in inverse proportion to the sine of the angles at which the planes are inclined. One has derived, therefore, the law of the inclined plane, or, equivalently, the resolution of forces into components. It is significant that this, probably the only significant discovery of the early Scientific Revolution concerning forces, is expressed as pure geometry.

Kinematics, the science of motion in itself without regard to its causes, is somewhat harder than statics to reduce to purely mathematical principles, but it can to some extent be done. The necessary connections between acceleration, speed and distance travelled are laid out in the simplest case by the medieval Merton Mean Speed Theorem, as described above, repeated by Galileo,[36] and cleared up in general by Newton and Leibniz's calculus. Feats such as Newton's derivation of Kepler's elliptical orbits from a central law of gravity are purely mathematical and need only principles of calculus.

It is otherwise with dynamics, which deals with the causes of motion such as gravity and other forces. It cannot be established *a priori* that the fall of heavy bodies is uniformly accelerated, much less what the constant of acceleration is. (That is why the contingency of the value of the gravitational constant and other constants of nature plays a crucial role in contemporary debates on the Anthropic Principle.) Galileo's observations were necessary to find the law of fall of heavy bodies.

Nevertheless, Galileo was at the same time responsible for one of the most remarkable demonstrations of the power of *a priori* mathematical reasoning in dynamics. When first considering what law should be followed by falling heavy bodies, once it is accepted that they go faster as they fall, he wondered how to distinguish between the two simplest theories: the perhaps most natural one that speed is proportional to the distance travelled from the start, and the equally simple but perhaps less natural one that speed is proportional to time from the start (that is, the body is uniformly accelerated, which is the correct answer).

Galileo realized, and was able to demonstrate, that the first theory needs no observations to refute it. It is absolutely impossible that acceleration should be proportional to the distance travelled. Galileo argues thus:

> When speeds have the same ratio as the spaces passed or to be passed, those spaces come to be passed in equal times; if therefore the speeds with which the falling body passed the space of four braccia were the doubles of the speeds with which it passed the first two braccia, as one space is double the other space, then the times of those passages are equal; but for the same moveable to pass the four braccia and the two in the same time cannot take place except in instantaneous motion.[37]

That reasoning is less than totally clear. But it suggests a diagram, surely present in some form in Galileo's mind, which clarifies it.

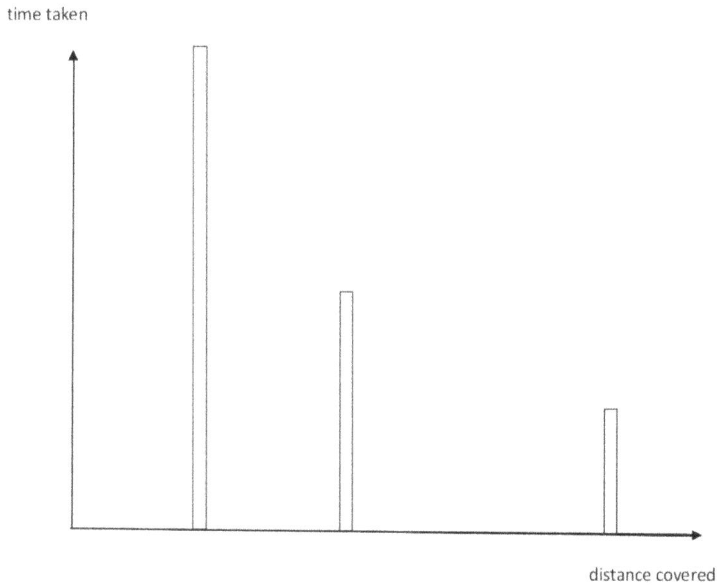

Figure 11.2 Galileo's proof that speed cannot be proportional to distance

The diagram should be read from right to left. If the time taken to cover a small distance is represented by the column of time above it (right column) then, at half that distance, it needs twice the time, at a quarter the distance, four times, and so on. As distance approaches zero, it needs indefinitely more time to cover the same interval. So, the motion can never get started.

From the falsity of the theory of the proportionality of speed to distance there does not follow, of course, the truth of the (true) alternative theory of the proportionality to time. But it leaves that theory as the natural simple alternative, guiding the effort of empirical confirmation.

Pascal's 'Geometry of Chance'

Probability is surely among the hardest of quantities for which to establish principles. It was far from clear that it was a quantity, in the sense of something to which a precise number could be assigned: surely the realm of uncertainty and chance of its nature resists being pinned down with precision? If we consider just the phenomena of games with dice and coins, the subject of Fermat and Pascal's 1654 correspondence that began the mathematical theory of probability, it is quite unclear what the quantity is to which a number should be assigned – one common modern answer, the long-run relative frequency of outcomes, was not considered in Fermat and Pascal's work nor in any other of that century.

Thus, the initial problem considered, very unnaturally from a later point of view, was a moral one: that of the just division of the stake in an interrupted game of chance. The quantity to be calculated is a moral right. Initially, there was a complete lack of principles or calculations, and no apparent place to start to find principles. Only Pascal realized that one might establish a realm of mathematical certainty in a new and unlikely field, the fluctuating realm of uncertainty. His comments on 'the astonishing title: the Geometry of chance (*aleae Geometria*)' were quoted in a previous chapter.[38]

Where might the principles be found? Pascal reaches for that old staple that had been so successful for Archimedes, Stevin and others in introducing *a priori* reasoning into applied mathematics, symmetry:

> The first principle for discovering how one should make the division is this. If one of the players is in a situation such that, whatever happens, a certain sum will belong to him in case of either a loss or a win, without chance being able to take it away from him, then he should make no division of it, but take it all, as assured . . .
>
> The second is this. If two players are in a situation such that, if one wins, a certain sum belongs to him, and if he loses, it belongs to the other; and if the game is of pure chance and there are as many chances for one as for the other, and consequently no more reason why one should win rather than the other, then if they wish to separate without playing, and take what legitimately belongs to each, the division is that they divide the sum that is at stake in halves, and take one each.[39]

Soon Huygens, writing the first book on probability and establishing it as a coherent body of theory, also laid down a symmetry principle as the basis for all calculations:

> Although in games determined solely by chance the outcomes are uncertain, there is always a fixed value for how much one has for winning over losing . . . I

> take it as a foundation that in a game the chance that one has towards something is to be estimated as such that, if one had it, one could procure the same chance in an equitable game. For example, if someone hides from me 3 shillings in one hand and 7 in the other, and gives me the choice of taking either hand, I say this is worth the same to me as if I were given 5 shillings. For if I have 5 shillings, I can again arrive at having an equal chance of getting 3 or 7 shillings, and that by an equitable game.[40]

The ability of *a priori* symmetry arguments to establish absolutely certain principles even in such an unlikely subject-matter as pure chance was the most perfect demonstration of their power.

The Aristotelian-Euclidean background to seventeenth-century mathematics means that mathematics does not fit at all into the old picture of the Scientific Revolution as a revolt against scholastic obfuscation in favour of empiricism and experimentalism. That model, even if over-simplified, does make some sense in some of the natural sciences, especially chemistry, biology and the more empirical parts of physics. Mathematics, including applied mathematics, is otherwise. The story of early modern applied mathematics is more like an extension of scholasticism than a retreat from it. Where the old scholastics themselves had been excessively modest about the possibilities of reducing the contingent physical world to quantitative order and demonstration, the mathematicians showed it could be done by doing it. Early modern applied mathematics is the pursuit of the scholastic vision of absolute necessities by other means.

The Survival of Aristotelian Science

Mathematico-Aristotelian science has not disappeared in the three centuries since then, although a philosophical view of it has been obscured by a succession of wrong or at best one-sided philosophies – both of science (inductivist, hypothetico-deductivist, falsificationist, Bayesian, and sociological), and of mathematics (nominalist, Platonist). While those philosophies of science emphasized the relation of uncertain theories to particular contingent facts, real sciences became increasingly mathematized, exposing the necessary interconnections between parts of scientific theory, even if not always establishing, as Aristotle had hoped, certain first principles themselves.

As we saw in Chapter 2, one of the central examples of real-world mathematical necessity, Euler's proof that the bridges of Königsberg could not be crossed once and once only, was developed in the eighteenth century. In the same period, d'Alembert promoted an almost entirely anti-experimental view of science, writing:

> No theory could have allowed us to find the law that heavy bodies follow in their vertical fall, but once this law is found through experience, all that belongs to the movement of heavy bodies, whether rectilinear or curvilinear, whether inclined or vertical, is found entirely by theory.[41]

D'Alembert's view has been vindicated by the course of theoretical physics since then, including conceptually driven science such as Einstein's thought experiment on simultaneity and the wide use of symmetry principles. Fluid dynamics, including weather prediction, is almost entirely mathematical with only some simple empirical physical principles needed. An example, the geometrical explanation of the Coriolis effect, was given in Chapter 4.

In the last hundred years, the realm of demonstrable mathematical necessities has expanded beyond what would normally be called applied mathematics to include a whole suite of completely new sciences, called the formal sciences or sciences of complexity.

The Formal Sciences Discover the Philosophers' Stone

Little noticed by philosophers of science, the 'formal sciences' such as operations research, theoretical computer science, systems engineering and statistics have flourished since the 1930s. They are wholly mathematical and have discovered the 'philosophers' stone', 'a way of transmuting opinion about the base and contingent beings of this world into the necessary knowledge of pure reason'.[42]

It used to be that the classification of sciences was clear. There were natural sciences, and there were social sciences. Then there were mathematics and logic, which may or may not be described as sciences, but seemed to be plainly distinguished from the other sciences by their use of proof instead of experiment and confirmation of hypotheses. That neat picture has been disturbed by the appearance in the last ninety years of a number of new sciences, variously called the 'formal' or 'mathematical' sciences, or the 'sciences of complexity'[43] or 'sciences of the artificial'.[44]

A paradigmatic formal science is operations research. Its origin is normally dated to the years just before and during World War II, when multi-disciplinary scientific teams investigated the most efficient patterns of search for U-boats, the optimal size of convoys, and the like.[45] Typical problems now considered are task scheduling, bin packing and timetabling. Given a number of factory tasks, subject to constraints about which must follow which, which cannot be run simultaneously because they use the same machine, and so on, one seeks the way to fit them into the shortest time. Bin packing deals with how to fit a heap of articles of given sizes most efficiently into a number of bins of given capacities.[46] The methods used rely essentially on search through the possibilities, using mathematical ideas to cut down the search space. Although it began in applications, OR is now very much an abstract science.

Many of the formal sciences have been driven by the computer revolution, which allows the computationally complex problems in those sciences to be solved. Theoretical computer science, studying such questions as the inherent computational complexity of algorithms and proofs of program correctness, stands as one of the most central, most mathematical and most applicable of the formal sciences. One significant part of it is the project of proof of correctness of software, or 'program verification', discussed in Chapter 4 above.

The list of such sciences is long including, but not limited to, control theory, signal processing, systems engineering, statistics and data science, network analysis, numerical taxonomy, information theory, mathematical ecology, game theory and (combining many of the forgoing) artificial intelligence. They all proceed by pure mathematical thought, assisted by extensive computer simulation and computation, and deliver real-world applicable results, results provable (in principle, sometimes in practice) with certainty.

Aristotelian science has returned.

Notes

1 James Franklin, *The Science of Conjecture: Evidence and Probability Before Pascal* (Baltimore: Johns Hopkins University Press, 2001/2015), 5–7.

2 Franklin, *Science of Conjecture,* 8–9.

3 Franklin, *Science of Conjecture*, 6, 9; Nachum L. Rabinovitch, *Probability and Statistical Inference in Ancient and Medieval Jewish Literature* (Toronto: University of Toronto Press, 1973), 112–15.

4 Azo of Bologna, c. 1200, quoted in Franklin, *Science of Conjecture,* 18.

5 *Corpus Iuris Canonici*, quoted in Franklin, *Science of Conjecture,* 23.

6 Franklin, *Science of Conjecture,* 32.

7 Franklin, *Science of Conjecture,* 28–33, 43–6.

8 Franklin, *Science of Conjecture*, chs 6–7; James Franklin, 'Late Scholastic Probable Arguments and Their Contrast with Rhetorical and Demonstrative Arguments', *Philosophical Inquiries* 10, no. 2 (2022).

9 Franklin, *Science of Conjecture*, ch. 4; Robert A. Maryks, *Saint Cicero and the Jesuits: The Influence of the Liberal Arts on the Adoption of Moral Probabilism* (London: Routledge, 2008); Rudolf Schuessler, *The Debate on Probable Opinions in the Scholastic Tradition* (Leiden: Brill, 2019).

10 Franklin, *Science of Conjecture*, chs 10–11; Giovanni Ceccarelli, *Risky Markets: Marine Insurance in Renaissance Florence* (Leiden: Brill, 2020).

11 James Franklin, 'Science by Conceptual Analysis: The Genius of the Late Scholastics', *Studia Neoaristotelica* 9 (2012): 3–24.

12 Anneliese Maier, *On the Threshold of Exact Science: Selected Writings of Anneliese Maier on Late Medieval Natural Philosophy* (Philadelphia: University of Pennsylvania Press, 1982); Edith D. Sylla, *The Oxford Calculators and the Mathematics of Motion, 1320–1350: Physics and Measurement by Latitudes* (Cambridge, MA: Harvard University Press, 1970); George Molland, *Mathematics and the Medieval Ancestry of Physics* (Aldershot: Variorum,1995); survey in Edith Sylla, 'The Oxford Calculators', in *The Cambridge History of Later Medieval Philosophy*, ed. N. Kretzmann, A. Kenny and J. Pinborg (Cambridge: Cambridge University Press, 1982), ch. 27; selections in Edward Grant, *A Source Book in Medieval Science* (Cambridge, MA: Harvard University Press, 1974), especially §§26, 41, 51.

13 James Hannam, *God's Philosophers: How the Medieval World Laid the Foundations of Modern Science* (London: Icon, 2009), chs 11–12.

14 Anneliese Maier, *Die Vorläufer Galileis im 14. Jahrhundert: Studien zur Naturphilosophie der Spätscholastik* (Rome: Edizioni di Storia e Letteratura, 1949).

15 Carl B. Boyer, *The History of the Calculus and its Conceptual Development* (New York: Dover, 1959), ch. 3.

16 Grant, *Source Book*, §§ 42–43.

17 William A. Wallace, *Domingo de Soto and the Early Galileo: Essays on Intellectual History* (Aldershot: Ashgate, 2004).

18 William A. Wallace, *Galileo's Early Notebooks: The Physical Questions* (Notre Dame, IND: University of Notre Dame Press, 1977); William A. Wallace, *Galileo and His Sources* (Princeton NJ: Princeton University Press, 1984); Eric A. Reitan, 'Thomistic Natural Philosophy and the Scientific Revolution', *Modern Schoolman* 73 (1996): 265–81.

19 Hannam, *God's Philosophers*, 330–2.

20 Norman Kretzmann, *Infinity and Continuity in Ancient and Medieval Thought* (Ithaca, NY: Cornell University Press, 1982).

21 Albert of Saxony, *Commentary on De Caelo*, summarized in Jacques Sesiano, 'On an Algorithm for the Approximation of Surds from a Provençal Treatise', in

Mathematics from Manuscript to Print 1300–1600, ed. C. Hay (Oxford: Clarendon, 1988), 30–56, at 42–7.

22 Odd Langholm, *Economics in the Medieval Schools* (Leiden: Brill, 1992); Odd Langholm, *The Legacy of Scholasticism in Economic Thought* (Cambridge: Cambridge University Press, 1998); Barry J. Gordon, *Economic Analysis Before Adam Smith: Hesiod to Lessius* (London: Macmillan, 1975), chs 6–9; John T. Noonan, *The Scholastic Analysis of Usury* (Cambridge, MA: Harvard University Press, 1957); Marjorie Grice-Hutchinson, *The School of Salamanca: Readings in Spanish Monetary Theory, 1544–1605* (Oxford: Clarendon, 1952); Raymond de Roover, 'The Scholastics, Usury and Foreign Exchange', *Business History Review* 41 (1967): 257–71; Joseph A. Schumpeter, *History of Economic Analysis* (London: Allen & Unwin, 1955), ch. 2; Jacob Viner, 'The Economic Doctrines of the Scholastics', *History of Political Economy* 10 (1978): 46–113; León Gómez Rivas, 'Business Ethics and the History of Economics in Spain: "The School of Salamanca": a Bibliography', *Journal of Business Ethics* 22 (1999): 191–202; Alejandro Antonio Chafuen, *Christians for Freedom: Late Scholastic Economics* (San Francisco: Ignatius Press, 1986); Wim Decock, 'Lessius and the Breakdown of the Scholastic Paradigm', *Journal of the History of Economic Thought* 31 (2009): 57–78.

23 Joel Kaye, *Economy and Nature in the Fourteenth Century: Money, Market Exchange and the Emergence of Scientific Thought* (New York: Cambridge University Press, 1998); Richard W. Hadden, *On the Shoulders of Merchants: Exchange and the Mathematical Conception of Nature* (Albany: SUNY Press, 1994), ch 4.

24 Hermann Schüling, *Bibliographie der psychologischen Literatur des 16. Jahrhunderts* (Hildesheim: Olms, 1967); Simon Kemp, *Medieval Psychology* (New York: Greenwood Press, 1990); Jack Zupko, 'What is the Science of the Soul?', *Synthese* 110 (1997): 297–334; Sven K. Knebel, 'Scotists vs. Thomists: What Seventeenth-Century Scholastic Psychology Was About', *Modern Schoolman* 74 (1997): 219–26.

25 Anthony Kenny, *Aquinas* (New York: Hill & Wang, 1980), ch 3; E. Ruth Harvey, *The Inward Wits: Psychological Theory in the Middle Ages and Renaissance* (London: Warburg Institute, 1975); Edward P. Mahoney, 'Sense, Intellect and Imagination in Albert, Thomas and Siger', in *The Cambridge History of Later Medieval Philosophy*, ch 30; Nicholas H. Steneck, 'Albert on the Psychology of Sense Perception', in *Albertus Magnus and the Sciences*, ed. J. A. Weisheipl (Toronto: Pontifical Institute of Mediaeval Studies, 1980), 263–90.

26 Stephen P. Stich, *From Folk Psychology to Cognitive Science* (Cambridge, MA: MIT Press, 1983); John D. Greenwood, ed, *The Future of Folk Psychology* (Cambridge: Cambridge University Press, 1991).

27 *Oxford English Dictionary*, article 'imagination'.

28 James Franklin, 'Diagrammatic Reasoning and Modelling in the Imagination: The Secret Weapons of the Scientific Revolution', in *1543 and All That: Image and Word,*

Change and Continuity in the Proto-Scientific Revolution, ed. G. Freeland & A. Corones (Kluwer, Dordrecht, 2000), 53–115.

29 Jerry Fodor, *The Modularity of Mind: An Essay in Faculty Psychology* (Cambridge, MA: MIT Press, 1983); Marvin Minsky, *Society of Mind* (New York: Simon & Schuster, 1986).

30 Stefan Meier-Oeser, 'Medieval Semiotics', *Stanford Encyclopedia of Philosophy,* 2003/2011; Edoardo Vineis and Alfonso Maierù, 'Medieval linguistics', in *History of Linguistics*, ed. G. Lepschy (London: Longman, 1994), vol 2, ch 2; G. L. Bursill-Hall, *Speculative Grammars of the Middle Ages,* The Hague: Mouton, 1971; Umberto Eco and Costantino Marmo, eds, *On the Medieval Theory of Signs* (Amsterdam: J. Benjamins, 1989); Ludger Kaczmarek, 'The Age of the Sign: New Light on the Role of the Fourteenth Century in the History of Semiotics', *Dialogue* (Canada) 31 (1992): 509–15; Claude Panaccio, 'Ockham and Locke on Mental Language', in *The Medieval Heritage in Early Modern Metaphysics and Modal Theory, 1400–1700*, ed. R. L. Friedman & L. O. Nielsen (Kluwer: Dordrecht, 2003), 37–51.

31 James Franklin, 'Mental Furniture from the Philosophers', *Et Cetera* 40 (1983): 177–91.

32 Peter M. Engelfriet, *Euclid in China: The Genesis of the First Chinese Translation of Euclid's Elements Books I–VI (Jihe yuanben; Beijing, 1607) and Its Reception Up to 1723* (Leiden: Brill, 1998).

33 Mary Domski, 'Observation and Mathematics', in the *Oxford Handbook of British Philosophy in the Seventeenth Century,* ed P.R. Anstey (Oxford: Oxford University Press, 2013), 144–68.

34 Simon Stevin, *De Beghinselen der Weeghconst* (Leiden, 1586), discussed in Josef T. Devreese and Guido Vanden Berghe, *'Magic Is No Magic': The Wonderful World of Simon Stevin* (Southampton: WIT Press, 2008).

35 Taken from http://en.wikipedia.org/wiki/Simon_Stevin⩾diaviewer/File:StevinEquilibrium.svg.

36 Boyer, *The History of the Calculus and its Conceptual Development,* 82–5.

37 Galileo, *Two New Sciences,* trans. S. Drake (Madison: University of Wisconsin Press, 1974), 160; John D. Norton and Bryan W. Roberts, 'Galileo's Refutation of the Speed-Distance Law of Fall Rehabilitated', *Centaurus* 54 (2012): 148–64.

38 Blaise Pascal, Address *Celeberrimae Matheseos Academiae Parisiensi* (1654), in *Oeuvres Complètes,* ed. J. Mesnard (Bruges: Desclée de Brouwer, 1964–70), vol. 2, 1034–5; Franklin, *Science of Conjecture,* 312.

49 Blaise Pascal, *Traité du Triangle Arithmétique,* Third Tract (1665), in *Oeuvres Complètes,* vol. 2, 1308; Franklin, *Science of Conjecture,* 311.

40 Christiaan Huygens, *De Ratiociniis in Ludo Aleae/Van Rekening in Spelen van Geluck* (On Reckoning in Games of Chance) (Leiden: 1657), in *Oeuvres Complètes de Christiaan Huygens* (La Haye: Nijhoff, 1888–) , vol. 14, 52–95; partial English

translation in Hans Freudenthal, "'Huygens' Foundations of Probability'", *Historia Mathematica* 7 (1980): 113–17; Franklin, *Science of Conjecture*, 314.

41 James Franklin, 'Artifice and the Natural World: Mathematics, Logic, Technology', in *Cambridge History of Eighteenth Century Philosophy*, ed. K. Haakonssen (Cambridge: Cambridge University Press, 2006), 817–53, section 3.

42 James Franklin, 'The Formal Sciences Discover the Philosophers' Stone', *Studies in History and Philosophy of Science*, 25 (1994), 513–33, updated as ch. 6 of *An Aristotelian Realist Philosophy of Mathematics*.

43 Heinz R. Pagels, *Dreams of Reason: The Computer and the Rise of the Sciences of Complexity* (New York: Simon & Schuster, 1988); M. Mitchell Waldrop, *Complexity: The Emerging Science at the Edge of Order and Chaos* (New York: Simon & Schuster, 1993); Per Bak, *How Nature Works: The Science of Self-organized Criticality* (New York: Copernicus, 1996); Stephen Wolfram, *A New Kind of Science* (Champaign, IL: Wolfram Media, 2002).

44 Herbert A. Simon, *The Sciences of the Artificial* (Cambridge, MA: MIT Press, 1969; 3rd edn, 1996).

45 Florence N. Trefethen, 'History of Operations Research', in *Operations Research for Management*, ed. J. F. McCloskey and F. N. Trefethen (Baltimore: Johns Hopkins Press, 1954), vol. 1, 3–35; S. I. Gass and A. A. Assad, eds, *An Annotated Timeline of Operations Research: An Informal History* (New York: Springer, 2004).

46 Robert E. D. Woolsey and Huntington S. Swanson, *Operations Research for Immediate Application: A Quick and Dirty Manual* (New York: Harper & Row, 1975); Hamdy A. Taha, *Operations Research: An Introduction*, 10th edn (Boston: Pearson, 2017).

12

Epilogue: How contact with the world of necessities was lost

Since the time of the ancient Greeks, many absolute necessities have been discovered beyond the narrow range of geometrical, logical and metaphysical ones known to Plato, Aristotle and Euclid. As we have seen in earlier chapters, the range of proved truths in pure and applied mathematics, logic and the formal sciences has vastly expanded, the necessary relations between evidence and conclusion have been understood, the interconnections in the more abstract empirical sciences like fundamental physics have become subject to mathematical proof, and the necessity of basic ethical truths has become clear.

Unfortunately, the more necessities that have been discovered, the less the prevailing philosophical culture has wanted to know about them. Since the Enlightenment, advances in real thought (and art, music and elsewhere) have been accompanied by, indeed swamped by, currents of thought that subject them to suspicion and resentment. Recalling a few highlights, or lowlights, of that story will explain the strength of the intellectual currents running against a straightforward acceptance of necessities. That will explain why the story of this book is out of touch with the spirit of the age.

At a cultural level, Anglophone nominalism in Baconian science, Locke's philosophy and English law's commitment to precedent over principle, its 'culture of fact'[1] valued 'experience' over theory. Despite Newton's *Mathematical Principles of Natural Philosophy* showing that even Englishmen could play at the top level of necessities, the empiricist way became the English way and was exported to the rest of the world as English became the international language. Anglophone philosophy fell into the orbit of Hume, whose division of knowledge into contingent matters of fact versus necessary 'relations of ideas' relegated necessity to the internal realm of thought. His baleful dictum 'no necessary connections between distinct existences'[2] left no room for necessities 'out there' in the external world.

The situation was exacerbated in the twentieth century when the three leading post-war American philosophers made different mistakes about modality that each distracted attention from genuine necessities – with Quine fudging the distinction between necessary and contingent,[3] Kripke focusing attention on alleged necessities following from naming such as 'Water is H_2O', and Lewis attempting to 'outsource' the necessities of this world to (mere) truth in an ensemble of all possible worlds.[4] Chapter 2 above described how those mistakes distracted attention from the absolute necessities that are the subject of this volume.

Continental thought from the Romantic era became captive to a different picture, which undermined necessities from another, more historicist, direction. Attention turned to the causes of belief in societies' core tenets (as opposed to the reasons for them), with the relativist implication that different societies with different histories might have different beliefs and so no beliefs are secure. That reasoning is an example of the argument identified by David Stove as 'the worst argument in the world' ('We can know things only as they are related to us/insofar as they fall under our conceptual schemes, etc., so, we cannot know things as they are in themselves')[5], but it has proved invaluable for those wishing to deconstruct any received ideas whatsoever. It fuelled the popular successes of the masters of the 'hermeneutic of suspicion'[6] such as Marx (with his class-based explanations of 'ideology') and Freud (whose explanations were psychoanalytically based). Objectivity in ethics was a particular target of those thinkers, being respectively put down to an imposture of the ruling class and repression. Their project has been followed by even more radicalism – and in academic circles more success – by recent followers like Foucault and Lacan. More recently, cause-for-reasons argumentation has driven evolutionary based attacks on the objectivity of ethics, such as the sociobiologist E.O. Wilson's argument that as the hypothalamus and limbic system flood the brain with all the emotions consulted by ethicists, and those organs are created by natural selection, that explains (away) 'ethics and ethical philosophers . . . at all depths'.[7] Anthropology and psychology have added their voices, aiming to achieve scientific status by explaining ethical behaviours without reference to ethics itself. The subsequent movement of thought loosely called postmodernism has gone even further in attacking the ability of mind and language to reach anything about reality, and has achieved success despite repeating the philosophical errors of earlier efforts.[8] The array of enemies of an intuitive human grasp of necessities has been not only numerous and diverse, but in the lead in the currents of humanistic thought in the last two centuries.

At the same time the higher levels of philosophy of ethics, which ought to have stood against popular currents, instead exacerbated them by arguing against the objectivity of ethics. John Mackie's *Ethics: Inventing Right and Wrong*,[9] which said there was no such thing as ethics on the grounds that it was too 'queer' a concept to fit into the scientific world picture, was just the most extreme point of a strong 'expressivist' or non-cognitivist tradition of metaethics which took ethics to express attitudes and commitments rather than advance truth claims.[10] If those less involved in the debate felt that was going too far, they remained uneasy about the apparent failure of other ethicists to reply to that school and explain where the objectivity of ethics could have come from. The unfoundedness of ethics was exported to the wider world of political theory by Rawls' immensely influential *Theory of Justice*, which used the ethical language of 'justice as fairness' but assured readers these concepts were 'political not metaphysical', that is, they were grounded in commitments rather than in reality.[11]

Meanwhile the understanding by non-mathematicians of what mathematics had achieved was undermined by strange interpretations of the alleged 'foundational crisis' in mathematics around 1900. While mathematicians showed no concern over this supposed crisis and regarded 'foundations' as an interesting but somewhat marginal subfield of pure mathematics, outsiders represented mathematics as beset by crisis and past its era of certainty.[12] At the same time, the discussion on foundations prompted philosophy of mathematics to diverge from anything mathematicians themselves could understand. Beginning with Frege and Russell's logicism, the philosophy of mathematics became an inward-looking field occupied by competing sects, with hard-line ('full-blooded'[13]) Platonists facing various nominalist schools such as logicism and formalism, which held that mathematics was not really about any subject matter but a mere language or method of deriving consequences from arbitrary axioms or manipulation of formal symbols. The domination of those views in philosophy hid from view everything that mathematicians were doing, especially any necessities arising in applied mathematics.[14] More recent philosophy of mathematics has failed to consider the Aristotelian alternative that would refocus attention on necessities. Philosophers bored with the conflict of traditional schools have instead pursued Lakatos's application to mathematics of the irrationalist programme in philosophy of science generally of the 'four modern irrationalists', Popper, Kuhn, Lakatos and Feyerabend.[15] Its aim was to overthrow the reliability of mathematical proof, a conclusion for which the humanities world of the 1970s and later had, needless to say, an enormous appetite. Equally obviously, it hid mathematical necessities behind a veil opaque to the humanist view. The gap between the 'two cultures', science and the humanities, was already wide in the mid-twentieth century. It was widened by the likes of the 'four irrationalists' giving permission for humanists to content themselves with a corrosive narrative about science and mathematics not grounded in the science and mathematics themselves.

Worse happened with the science of statistics. The twentieth century saw an extraordinary blossoming of statistical methods that allowed the rigorous testing of hypotheses in drug trials, safety science and many other areas. As with mathematics, quantum mechanics and so many other areas, these advances in rationality reached the humanistic world through a lens of misinterpretation. The leading explainer of the meaning of hypothesis testing, the statistician Jerzy Neyman, gave a frequentist interpretation of hypothesis testing, according to which it did not involve in any way reasoning or belief but only behaviour. As a result, the official story of statistics textbooks completely hid the necessary connections between evidence and conclusion.[16]

These sad developments were not confined to necessities. They can hardly be understood without putting them in the context of other areas where Western Civilization achieved a Golden Age of contact with the eternal but then lost it. Classical music from, say Monteverdi's *Vespers* through Vivaldi, Mozart and Beethoven to César Franck's *Panis angelicus*, leaves an impression that the composers of genius enjoyed direct inspiration from another world. A Silver Age followed, characterized by many works of genius but which lacked the same simplicity as earlier compositions. Eventually, it became challenging to produce classical music of the highest calibre, although a significant audience with a somewhat countercultural inclination continues to appreciate and listen to such music. Similar stories can be told of painting from Giotto to Vermeer and Velázquez, and of literature from Dante to Jane Austen. With effort and education, a contemporary person can tune in to those geniuses and understand their classic productions. But we cannot be them, and we cannot repeat their unselfconscious and unmediated access to the world of forms. Modern nervousness prevents it, just as it prevents us from accepting with untroubled hearts our access to the world of necessities.

At the dawn of the ancient Greek classical era, oracles were a fact of life. The most famous one, at Delphi, set Western philosophy on its course by assuring Socrates that he was the wisest man in Greece, leaving him to deduce how that was possible when he knew he knew nothing. Thereafter it was observed that the oracles slowly became harder to contact. Originally sites where the voices of the gods could be heard freely and without mediation, they gradually came to need priestesses in states of trance to give voice to the divine, then interpreters who could explain the increasingly garbled utterances of the priestesses. Finally, around the first century CE, the oracles fell silent.[17] Cultural conditions had made contact with divine voices impossible.

The ancient oracles were false. Our intuitions of necessity, on the contrary, are real and in general reliable oracles that we carry around with us and consult at will. The cultural veils and intellectual fallacies that obscure them must be removed. The oracles must be restored.

Notes

1 Barbara J. Shapiro, *A Culture of Fact: England 1550–1720* (Ithaca, NY: Cornell University Press, 1999); the survival of broadly Aristotelian assumptions described in Peter R. Anstey, 'Locke and the Problem of Necessity in Early Modern Philosophy', in *Logical Modalities from Aristotle to Carnap*, ed. Max Cresswell, Adriane Rini and Edwin Mares (Cambridge: Cambridge University Press, 2016), 174–93, section 1.

2 In a modern rendering, see Jessica Wilson, 'What is Hume's Dictum, and Why Believe It?', *Philosophy and Phenomenological Research* 80, no. 3 (2010): 595–637.

3 Roberta Ballarin, 'Quine on Modality', in *The Routledge Handbook of Modality,* ed. Otávio Bueno and Scott A. Shalkowski (Abingdon: Routledge, 2021), 390–99.

4 Scott A. Shalkowski, 'Modalism', in *The Routledge Handbook of Modality,* ed. Otávio Bueno and Scott A. Shalkowski (Abingdon: Routledge, 2021), 114–22, at 118.

5 James Franklin, 'Stove's Discovery of the Worst Argument in the World', *Philosophy* 77 (2002): 615–24.

6 Paul Ricoeur, *Freud and Philosophy: An Essay on Interpretation* (New Haven: Yale University Press, 1970); Alison Scott-Baumann, *Ricoeur and the Hermeneutics of Suspicion* (London: Bloomsbury, 2009).

7 E. O. Wilson, *Sociobiology: The New Synthesis* (Cambridge, MA: Belknap Press, 1975), 3; discussed in Franklin, 'Stove's Discovery'.

8 Well exposed in Raymond Tallis, *Not Saussure: A Critique of Post-Saussurean Literary Theory* (Basingstoke: Macmillan, 1988).

9 J. L. Mackie, *Ethics: Inventing Right and Wrong* (Harmondsworth: Penguin, 1977).

10 Mark van Roojen, 'Moral Cognitivism vs. Non-Cognitivism', *Stanford Encyclopedia of Philosophy,* 2004/2023.

11 John Rawls, 'Justice as Fairness: Political Not Metaphysical', *Philosophy & Public Affairs* 14, no. 3 (1985): 223–51 at 230–1; James Franklin, *The Worth of Persons: The Foundation of Ethics* (New York: Encounter Books, 2022), 62–8.

12 Morris Kline, *Mathematics: The Loss of Certainty* (New York: Oxford University Press, 1980); its many hostile reviews include James Franklin, 'Negativism', *Quadrant* 27, no. 8 (August 1983): 82–4.

13 Greg Restall, 'Just What Is Full-Blooded Platonism?', *Philosophia Mathematica* 11 (2003): 82–91.

14 Pointed out in Stephan Körner, *The Philosophy of Mathematics: An Introductory Essay* (London: Hutchinson, 1960), ch. 8.

15 Imre Lakatos, *Proofs and Refutations: The Logic of Mathematical Discovery* (Cambridge: Cambridge University Press, 1976); D. C. Stove, *Popper and After: Four Modern Irrationalists* (Oxford: Pergamon, 1982); James Franklin, 'Last Bastion of Reason', *New Criterion* 18, no. 9 (May 2000): 74–8.

16 James Franklin, 'Resurrecting Logical Probability', *Erkenntnis* 55 (2001): 277–305, section 7.

17 Plutarch, *De Defectu Oraculorum*; D. C. Stove, 'The Oracles and Their Cessation: A Tribute to Julian Jaynes', in David Stove, *Cricket Versus Republicanism: And Other Essays* (Sydney: Quakers Hill Press, 1995), 115–40.

Select Bibliography

[This is not in any sense a complete bibliography of work on absolute necessities or of works referred to. It lists just those books and articles that have proved especially valuable for forming the point of view of this book.]

Archimedes, *On the Equilibrium of Planes,* in *The Works of Archimedes,* ed T. L. Heath. Cambridge: University Press, 1897.

Aristotle, *Categories.*

Aristotle, *Posterior Analytics.*

Armstrong, D. M. *Universals and Scientific Realism*. Cambridge: Cambridge University Press, 1978.

Bibby, Neil. 'Tuning and Temperament: Closing the Spiral'. In *Music and Mathematics: From Pythagoras to Fractals*, ed. J. Fauvel, R. Flood and R. J. Wilson, 13–28. Oxford: Oxford University Press, 2003.

Bigelow John and Robert Pargetter. *Science and Necessity*. Cambridge: Cambridge University Press, 1990.

Boyer, Carl B. *The History of the Calculus and its Conceptual Development*. New York: Dover, 1959.

Bronstein, David. *Aristotle on Knowledge and Learning: The Posterior Analytics*. Oxford: Oxford University Press, 2016.

Brown, James Robert. *Philosophy of Mathematics: A Contemporary Introduction to the World of Proofs and Pictures,* 2nd edn, New York, Routledge, 2008.

Bueno, Otávio and Scott A. Shalkowski, ed. *The Routledge Handbook of Modality.* London and New York: Routledge, 2021.

Euclid, *Elements.*

Euler, Leonhard. 'Solutio problematis ad geometriam situs pertinentis' (1735), trans. in *Graph Theory 1736–1936*, ed. N. Biggs, E. Lloyd and R. Wilson, 3–8. Oxford: Oxford University Press, 1976.

Fine, Kit. 'The Varieties of Necessity'. In *Conceivability and Possibility,* ed. Tamar Gendler and John Hawthorne, 253–81. New York: Oxford University Press 2002.

Forrest, Peter and D. M. Armstrong. 'The Nature of Number'. *Philosophical Papers* 16 (1987): 165–86.

Gaita, Raimond. *Good and Evil: An Absolute Conception*. 2nd edn, Abingdon: Routledge, 2004.

Hoare, C. A. R. 'An Axiomatic Basis for Computer Programming'. *Communications of the Association for Computing Machinery* 12 (1969): 576–80.

Irvine, Andrew D. 'Frege on Number Properties'. *Studia Logica* 96 (2010): 239–60.
Jaynes, E. T. *Probability Theory: The Logic of Science*. Cambridge: Cambridge University Press, 2003.
Kant, Immanuel. *The Metaphysics of Morals* (1797), trans. M. Gregor. Cambridge: Cambridge University Press, 1991.
Kessler, Glenn. 'Frege, Mill and the Foundations of Arithmetic'. *Journal of Philosophy* 77 (1980): 65–79.
Keynes, J. Maynard. *Treatise on Probability*. London: Macmillan, 1921.
Leibniz, Gottfried. *Theodicy: Essays on the Goodness of God, the Freedom of Man and the Origin of Evil* (1710), trans. E. M. Huggard, ed. Austin M. Farrer. London: Routledge and Kegan Paul, 1951.
Lonergan, Bernard. *Insight: A Study of Human Understanding*. 5th edn, Toronto: Toronto University Press, 1997.
McKirahan, Richard D. *Principles and Proofs: Aristotle's Theory of Demonstrative Science*. Princeton: Princeton University Press, 1992.
Moore, G. E. *Principia Ethica*. Cambridge: Cambridge University Press, 1903.
Morreau, Michael. 'Arrow's Theorem'. *Stanford Encyclopedia of Philosophy* (2014/2019).
Norton, John D. and Bryan W. Roberts. 'Galileo's Refutation of the Speed-Distance Law of Fall Rehabilitated'. *Centaurus* 54 (2012): 148–64.
Pap, Arthur. 'Nominalism, Empiricism and Universals: I', *Philosophical Quarterly* 9 (1959): 330–40.
Polya, George. *Patterns of Plausible Inference*. 2nd edn, Princeton: Princeton University Press, 1968.
Shalkowski, Scott A. 'Logic and Absolute Necessity'. *Journal of Philosophy* 101 (2004), 55–82.
Sher, Gila. *Logical Consequence*. Cambridge: Cambridge University Press, 2022.
Stove, D. C. *The Rationality of Induction*. Oxford: Clarendon, 1986.
Stevin, Simon. *De Beghinselen der Weeghconst*. Leiden, 1586.
Studtmann, Paul. 'Aristotle's *Categories*'. *Stanford Encyclopedia of Philosophy*, 2021.
Tahko, Tuomas E. 'The Law of Non-Contradiction as a Metaphysical Principle'. *Australasian Journal of Logic* 7 (2009), 32–47.
Waldrop, M. Mitchell. *Complexity: The Emerging Science at the Edge of Order and Chaos*. New York: Simon & Schuster, 1993.
Weinberg, Julius R. *Abstraction, Relation, Induction*. Madison: University of Wisconsin Press, 1965.
Wilson, Jessica. 'What is Hume's Dictum, and Why Believe It?' *Philosophy and Phenomenological Research* 80 (2010), 595–637.
Williams, D. C. *The Ground of Induction*. Cambridge, MA: Harvard University Press, 1947.
Wood, Allen. *Kant's Ethical Thought*. New York: Cambridge University Press, 1999.

Index

www.ingramcontent.com/pod-product-compliance
Ingram Content Group UK Ltd.
Pitfield, Milton Keynes, MK11 3LW, UK
UKHW020817290926
15470UKWH00003B/1212

9 781350 467071